MATHEMATICAL CHALLENGES

SELECTED PROBLEMS FROM THE MATHEMATICS STUDENT JOURNAL

COMPILED AND ANNOTATED BY

MANNIS CHAROSH

NATIONAL COUNCIL OF TEACHERS OF MATHEMATICS

1201 SIXTEENTH STREET, N.W., WASHINGTON, D.C. 20036

Library of Congress Catalog Card Number: 65-19606

Printed in the United States of America

PREFACE

This collection of problems is addressed particularly to students in grades 7-12 and assumes no background beyond the eleventh year. Practically all of the problems have appeared in issues of the *Mathematics Student Journal* from the first issue of February, 1954. Original problems as well as some quoted from other sources are included. A small number of problems that have never appeared in the Journal have been added to illustrate certain useful concepts. The criteria for inclusion were: (1) that the problems be different from the routine types encountered in the classroom; (2) that they be interesting and challenging; (3) that they come within the scope of the intended grade level.

The problems are not completely graded with respect to difficulty. The chapter divisions are not precise and are not intended to be so. Thus it would be expected that techniques usually referred to as *algebraic, geometric,* or *trigonometric* may be needed in almost any problem in the book. This is as it should be, because the subject of these problems is *mathematics,* and the terms *geometry, algebra, trigonometry* merely indicate areas of emphasis.

Many of these problems have been solved at various levels. For some problems, students in lower grades often work by trial and error using little or no analysis. This method provides a useful base upon which to build a more profound understanding of mathematical concepts. Students in higher grades are able to apply more sophisticated analyses in their solutions.

In the section devoted to solutions, more than one solution may be given to exhibit a variety of techniques intended to broaden the mathematical experience of the student. Alternate solutions may require a background beyond the eleventh year. For the latter, explanations are given in the text of the solution, or in one of the appendices, or in the references.

The appendices give theorems and proofs that are usually not included within the scope of grades 7-12. The references are indicated in brackets: [ ] and may be found in the bibliography at the end of the book. Annotations may refer to historical references, alternate solutions, related problems, or other comments.

Problem solving is valuable experience because:

1. It provides the satisfaction that accompanies the accomplishment of a difficult task.
2. It provides the opportunity to develop mathematical skills by applying known principles, formulas, and theorems.
3. It enables the student to devise new approaches to mathematics and to begin research activity in mathematics.
4. It teaches the student to appreciate the possibilities of elegance in a solution, and on a larger scale, to appreciate the beauty in mathematics as a whole.

Finally a word to the solver on how to use this book.

1. Try to solve the problems of types that are familiar to you; then go to the other types.
2. If you fail to solve a problem, put it aside and come back to it later. Trying to solve a problem may be rewarding, even if success does not come at once.
3. After solving a problem, read the solution in the book and any annotations that may be given. You may find suggestions that you will wish to pursue.
4. If you are convinced that you cannot solve a problem, read the beginning of the solution in the book, then try to complete it yourself.
5. If you still cannot solve the problem, read the entire solution. You may find that you can then apply the ideas learned to other problems.

The writer wishes to thank Professor Myron F. Rosskopf of Columbia University for his valuable advice in the planning of this book.

MANNIS CHAROSH

Brooklyn, New York
October 1964

CONTENTS

SYMBOLS USED IN THIS BOOK

Solutions received from every part of the country have indicated that more and more schools are beginning to use the symbols found in the geometry textbooks prepared by the School Mathematics Study Group. Therefore these symbols will be used in this book. They are as follows:

$\overleftrightarrow{AB}$	The line containing the two points A and B
$\overline{AB}$	The segment having A and B as end-points
$\overrightarrow{AB}$	The ray with A as its end-point and containing point B
$\angle ABC$	The angle having B as vertex and $\overrightarrow{BA}$ and $\overrightarrow{BC}$ as sides
$\triangle ABC$	The triangle having A, B, and C as vertices
AB	The positive number which is the distance between the two points A and B, and also the length of the segment $\overline{AB}$
$m\angle ABC$	The real number between 0 and 180 which is the degree measure of $\angle ABC$
$\cong$	**Congruence.** $\overline{AB} \cong \overline{CD}$ means that two segments (for example sides of a triangle) treated as sets of points are congruent. Similarly for $\angle ABC \cong \angle DEF$
$=$	**Equality.** $AB = CD$ means that the lengths of segments AB and CD are equal. Similarly for $m\angle ABC = m\angle DEF$.

I

A MISCELLANY OF BREVITIES

1. If a clock strikes six times in five seconds, how many times will it strike in ten seconds?

2. What is the least positive integer by which 180 should be multiplied, so that the product is **(a)** a perfect square; **(b)** a perfect cube?

3. It takes 852 digits to number the pages of a book consecutively.

(a) How many pages are there?

(b) How many times is the digit 7 printed?

4. A man bought a horse for \$60, sold it for \$70, then bought it back for \$80, and finally sold it for \$90. How much money did he make or lose in this series of transactions?

5. Find the smallest fraction that each of these will divide exactly:

$$\frac{6}{7}, \frac{5}{14}, \frac{10}{21}.$$

6. If, in a single purchase, you are offered 3 successive discounts of 20 percent, 10 percent, 5 percent in any order that you wish, what order would you choose?

7. A man's salary is reduced by p percent. By what percent would his salary then have to be raised to bring it back to the original amount?

8. For what must a merchant mark goods costing him c dollars, so that he can sell them at a discount of d percent of the marked price, and still make a profit of d percent of the selling price?

9. **(a)** If a man drives to work at the rate of 15 miles per hour and returns along the same route at the rate of 30 miles per hour, what is his average rate for the entire round trip?

(b) Do the same problem replacing 15 and 30 by x and y respectively.

10. In the series $+1\ -1\ +1\ -1\ +\ldots$, in which $+1$'s and -1's alternate, find a formula for the nth term.

11. Write the quadratic equation whose roots are the reciprocals of those of **(a)** $6x^2-11x+4=0$; **(b)** $ax^2+bx+c=0$.

12. Find the value of **(a)** $b^{\log a}$; **(b)** $b^{-\log a}$. The base of the logarithm in each case is b.

13. If $i=\sqrt{-1}$ and $i^r=i^s$, what relationship exists between r and s?

14. Find the value of $(i^9+i^{10}+i^{11}+i^{12})^3$. Generalize.

15. Solve for x: $8^{\frac{1}{6}}+x^{\frac{1}{3}}=\dfrac{7}{3-\sqrt{2}}$.

16. **(a)** If N is an integral square, what is the next consecutive integral square?

(b) If N is an integral cube, what is the next consecutive integral cube?

17. Solve for x: $\dfrac{\sqrt{x+1}+\sqrt{x-1}}{\sqrt{x+1}-\sqrt{x-1}}=3$.

18. Three mutually tangent circles have centers A, B, C and radii a, b, c respectively. The lengths of segments AB, BC, CA are 17, 23, and 12 respectively. Find the lengths of the radii.

19. In Fig. 1, find the sum: $m\angle A+m\angle B+m\angle C+m\angle D+m\angle E$.

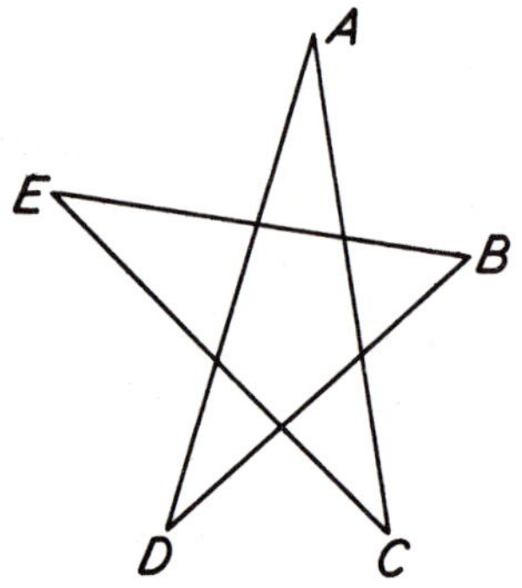

Figure 1

20. $ABCD$ is a trapezoid, with $\overline{AB}\,||\,\overline{CD}$. E and F are taken on $\overline{AD}$ and $\overline{BC}$ respectively, so that $\dfrac{AE}{ED}=\dfrac{BF}{FC}=\dfrac{2}{1}$. If $AB=7$, and $DC=10$, find the length of EF.

21. A square piece of tin is made into an open box by cutting from each corner a square of side *s* inches and then turning up the sides. If the box is to contain *c* cubic inches, find the dimensions of the original piece of tin.

22. Represent the number *one* by using **(a)** just three *nines* and a minus sign; **(b)** one *nine* and two dots.

23. Show that no pair of integers satisfies the equation $3x+6y=95$.

24. There are eight men in a room. Each one shakes hands with each of the others once. How many handshakes are there?

25. In each of the following, each letter stands for a single digit. The codes are independent of each other. What does each letter stand for?

(a) $Y+Y+Y=MY$.

(b) $XXX+B=BAAA$.

(c) $MA+A=AM$.

(d) $ON+ON+ON+ON=GO$.

26. Restore the missing digits in this exact division.

```
         9 x x
3 x ) x 4 x 9 x
      x 1 x
      -----
        x x x
        x 1 x
        -----
          2 x 5
          2 x 5
          -----
```

27. Bill, John, Henry, and Joe have to catch the 6 o'clock train.

(1) Bill's watch is ten minutes fast, but he thinks that it's five minutes slow.

(2) John's watch is ten minutes slow, but he thinks that it's ten minutes fast.

(3) Joe's watch is five minutes slow, but he thinks that it's ten minutes fast.

(4) Henry's watch is five minutes fast, but he thinks that it's ten minutes slow.

Each of them leaves to catch the train so that he will just make it if his time is what he thinks it is. Who misses the train?

28. Fill in the blank spaces in the magic square shown in Fig. 2 using each of the following numbers: 4, 6, 7, 8, 10, 11, 12, 13, 14, 15, so that the sum of the integers in each column, row, and diagonal is 34.

	1		
	5		
2	16	9	
3			

Figure 2

II

PROBLEMS WITH A GEOMETRIC BACKGROUND

29. A running track has straight parallel sides and semicircular ends, and it measures y yards along the inside curb. The lanes are numbered outward from this curb, and each is a yard wide. How much headstart should a runner in the second lane receive over a runner in the first lane, if each is to cover a distance of y yards by the time he reaches the finish line?

30. In Fig. 3: $\overline{AB} \| \overline{FE} \| \overline{DC}$. Prove: $\frac{1}{EF} = \frac{1}{AB} + \frac{1}{CD}$.

Figure 3

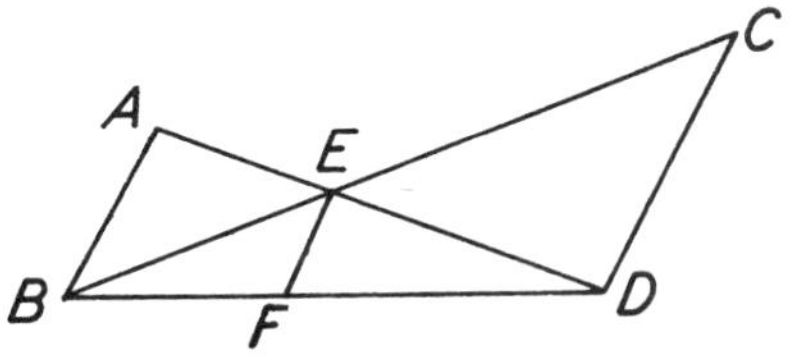

31. I is the center of the inscribed circle of $\triangle ABC$. $\overrightarrow{AI}$, extended, intersects the circumference of the circumscribed circle at D. Prove that D is equidistant from I, B, and C.

32. If the length of a side of a rhombus is the mean proportional between the lengths of its diagonals, find the number of degrees in the acute angle between its sides.

33. Given a right triangle whose hypotenuse and altitude to the hypotenuse have lengths c and h respectively:

(a) Express in terms of c and h, the length of a side of the inscribed square having two of its sides resting on the legs of the triangle.

(b) Express in terms of c and h, the length of a side of the inscribed square resting on the hypotenuse.

(c) Which is the larger square?

34. $ABCD$ is a trapezoid with upper base $AB=a$ and lower base $DC=b$. M is taken on $\overline{AD}$ and N on $\overline{BC}$ so that $\overline{MN}\|\overline{AB}$ and $\overline{MN}$ bisects the area of the trapezoid. Express the length x of $\overline{MN}$ as a function of a and b.

35. Construct a circle that passes through two given points and is tangent to a given line. Is this possible for any line and any two points?

36. $ABCD$ is a parallelogram and E is a point on $\overrightarrow{CD}$ prolonged. $\overline{BE}$ is drawn, intersecting $\overline{AD}$ at F and diagonal $\overline{AC}$ at G.

(a) Find BG if $BF=3$ and $BE=6$; $BF=10$ and $BE=15$; $BF=7$ and $BE=8$.

(b) What generalization do the results suggest? Prove it.

37. Triangle ABC is an isosceles right triangle with hypotenuse $\overline{AB}$. $\overline{CD}$ is drawn parallel to $\overline{AB}$ so that $BD=BA$. Find the number of degrees in $m\angle DBC$ for each of the two possible positions of D.

38. A square 12 in. by 12 in. is divided into n^2 equal squares by equally spaced lines parallel to its sides; and circles are inscribed in the squares.

(a) Find the sum of the areas of the circles for $n=1$, 2, 3, and 4.

(b) What generalization do the results suggest? Prove it.

39. The sides of $\triangle ABC$ (Fig. 4) are $AB=6$, $BC=5$, $CA=8$. At point D on $\overline{AB}$, a line is drawn parallel to $\overline{AC}$, intersecting $\overline{BC}$ at E; then

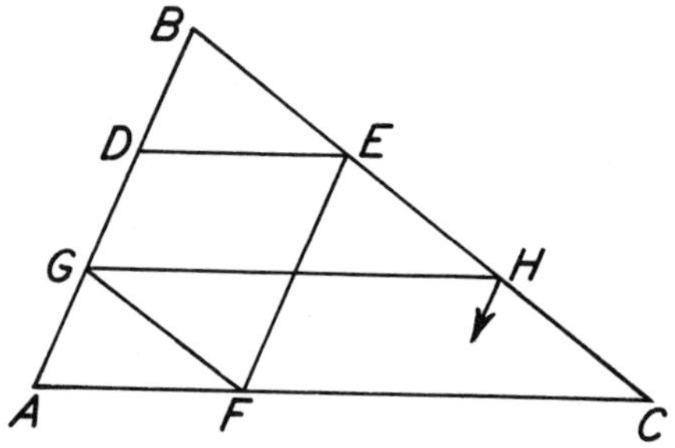

Figure 4

parallel to $\overline{AB}$, intersecting $\overline{AC}$ at F; then parallel to $\overline{BC}$, intersecting $\overline{AB}$ at G; then parallel to $\overline{AC}$ intersecting $\overline{BC}$ at H; then parallel to $\overline{AB}$, intersecting $\overline{AC}$ at I; then parallel to $\overline{BC}$, intersecting $\overline{AB}$ at J.

(a) Find AG and BJ, when $BD=2$; 3; 5.

(b) Generalize and prove your results.

(c) Under what conditions will $\overline{FE}$, $\overline{GH}$, and $\overline{DI}$ be concurrent?

40. The inscribed circle of right triangle ABC is tangent to hypotenuse $\overline{AB}$ at D.

(a) Find the area of the triangle if $AD=2$, $DB=3$; $AD=3$, $DB=10$.

(b) What generalization do the results suggest? Prove it.

41. One leg of a right triangle has a length of 48, and the other two sides have integral lengths. Find the lengths of the other two sides.

42. In quadrilateral $ABCD$, BA and BC are each equal to t, and $m\angle ADC = m\angle A + m\angle C$. Express BD in terms of t.

43. In $\triangle ABC$, $m\angle C = 30°$; $\triangle ABD$ is equilateral with D and C on opposite sides of $\overleftrightarrow{AB}$. Prove that $\overline{CA}$, $\overline{CB}$, $\overline{CD}$ can be the sides of a right triangle. Consider the cases where $m\angle CAB$ is less than, equal to, or greater than $120°$

44. **(a)** Prove that if $ABCD$ is a parallelogram and if the altitude from A to $\overleftrightarrow{DC}$ and from A to $\overleftrightarrow{BC}$ have equal lengths, then the figure is a rhombus.

(b) Apply (**a**) to construct a rhombus so that two of its sides lie on two given parallel lines l_1 and l_2, and the other two sides pass through two given points, P and Q.

45. Construct a right triangle, given the lengths of the hypotenuse and the sum of its legs.

46. A diameter $\overline{AB}$ of a certain circle O is extended through A to C so that $AC = 3''$. At B, $\overline{BD}$ is drawn perpendicular to $\overline{AB}$ so that $BD = 9''$. Determine length AB so that $\overline{DC}$ shall be tangent to the circle.

47. $\overline{AD}$ is a diameter of circle O. Quadrilateral $ABCD$ is inscribed so that AB, BC, $CD = a$, b, c, respectively. If the length of the diameter is x, find an equation whose solution will give x in terms of a, b, c.

48. Devise a method of trisecting a line segment, using only straightedge and compasses, but without use of parallel lines.

49. In Fig. 5, O is the center of the circle, and C and D trisect chord $\overline{AB}$. Prove that the measures of angles 1, 2, 3 cannot all be equal.

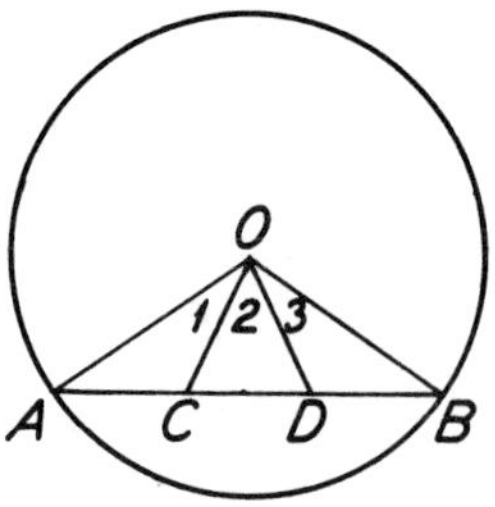

Figure 5

50. $ABCD$ is a square; F is the midpoint of $\overline{DC}$ and E is taken on $\overline{BF}$ so that $\overline{AE}$ is perpendicular to $\overline{BF}$. Prove that $AD = DE$. Is this true for any rectangle? For any parallelogram?

51. In $\triangle ABC$, with $m\angle ACB = 120°$, and $\overline{CD}$ the bisector of $\angle ACB$ with D on $\overline{AB}$, prove that $\frac{1}{CD} = \frac{1}{AC} + \frac{1}{BC}$.

52. Two squares (Fig. 6), each with side 12, are placed so that a corner of one lies at the center of the other.

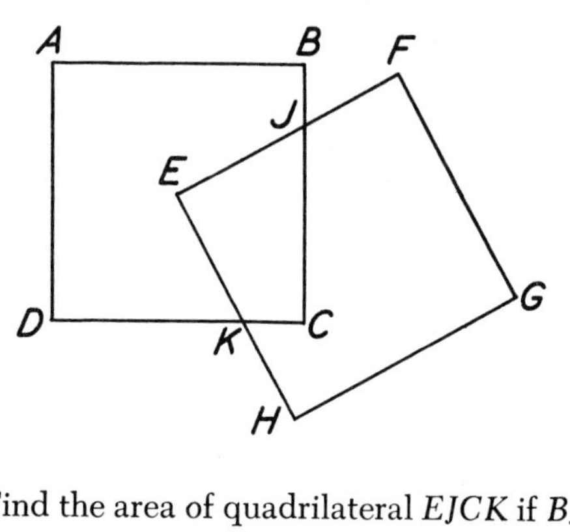

Figure 6

(a) Find the area of quadrilateral $EJCK$ if $BJ = 3$; 4; 6.

(b) What generalization do your results suggest? Prove it.

53. Draw any radius $\overline{OA}$ of circle O and its perpendicular bisector $\overline{LK}$, with L on $\overline{OA}$ and K on the circumference. If chords of length equal to LK are laid off consecutively around circle O, it will seem that a regular heptagon (seven-sided polygon) has been constructed. Is this construction correct?

54. Point D is taken on side $\overline{BC}$ of $\triangle ABC$ so that if circles are inscribed in each of the triangles ABD and ACD, these circles will be tangent to each other at a point on $\overline{AD}$. The lengths AB, BC, CA, and BD are represented by c, a, b, and x, respectively.

(a) Find x in terms of a, b, and c.

(b) If the inscribed circle of $\triangle ABC$ is tangent to $\overline{BC}$ at E, find the length DE.

55. In Fig. 7 quadilateral $ABCD$ is inscribed in a circle. $AE=AD$; $\overline{AE}\,||\,\overline{BC}$; $CF=CD$; $\overline{CF}\,||\,\overline{AB}$. Prove that $\overline{DF}\perp\overline{DE}$.

Figure 7

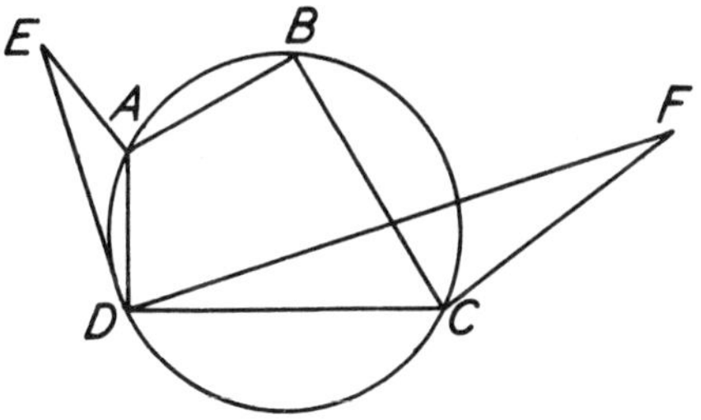

56. Prove that if the vertices of a triangle have coordinates that are integers, the triangle cannot be equilateral.

57. $AEDC$ and $BCFG$ are squares on the legs of right triangle ABC. Other lines are drawn as in Fig. 8. Prove that the area of triangular region AIB equals the area of quadrilateral $CHIJ$.

Figure 8

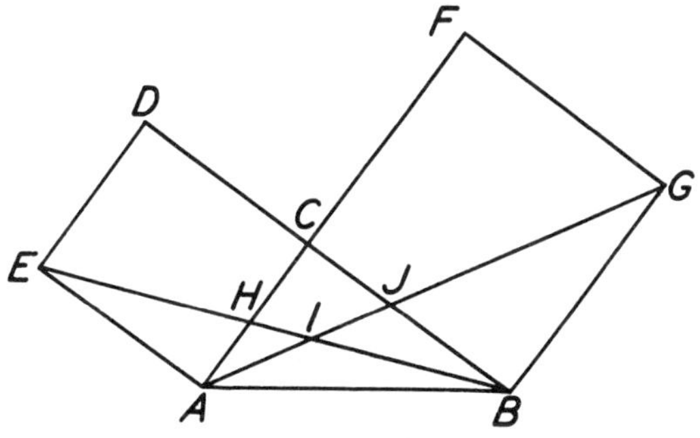

58. In Fig. 9 $\overline{AB}$ is tangent to the circle at B; D is any point on the circle, and E is any point such that $AB=AE$. $\overline{AD}$ and $\overline{ED}$ intersect the circle at C and G respectively. $\overline{EC}$ intersects the circle at F. Prove: $\overline{FG}||\overline{AE}$. May E be taken inside the circle?

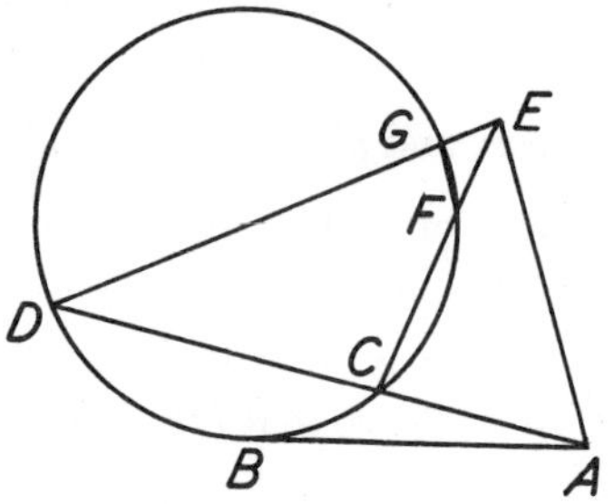

Figure 9

59. E is taken inside square $ABCD$ so that $m\angle EAD=m\angle EDA=15°$ Prove that $\triangle EBC$ is equilateral.

60. $ABCDEF$ is a regular hexagon. G and H are midpoints of $\overline{AB}$ and $\overline{CD}$ respectively. $\overline{EG}$ intersects $\overline{AD}$ at P, and $\overline{FH}$ intersects $\overline{AD}$ at Q. Prove that $AP=PQ=QD$.

61. Semicircles are erected outwardly on the three sides of right triangle ABC, using the sides as diameters. Quadrilateral $EFGH$ is circumscribed around the figure so that $\overline{EF}$ and $\overline{GH}$ are parallel to the leg $\overline{BC}$ and also tangent to their respective semicircles, and $\overline{FG}$ and $\overline{HE}$ are parallel to the leg $\overline{AC}$ and also tangent to their respective semicircles. Prove that the quadrilateral $EFGH$ is a square.

62. P is a point inside square $ABCD$ such that $PA=5$, $PB=3$, and $PC=7$. Find the length of the side of the square.

63. Prove that it is possible for two right triangles to have five pairs of equal parts (sides, angles) and not be congruent.

64. A man seeks to conceal treasure using as landmarks an elm tree E, a pine tree P which is a little further west, and a gallows G in the vicinity of and north of the two trees. He locates two points A and B as follows: To locate A, proceed from G to P, turn right 90°, and continue to A so that $GP=PA$. To locate B, proceed from G to E, turn left 90° and continue to B so that $GE=EB$. He then buries his treasure at M, the midpoint of $\overline{AB}$. When he returns some months later to recover his hoard, he is shocked to discover that the gallows has been removed so that no trace remains. Show that it is still possible for him to locate his treasure.

65. Show how to construct through a given point a line which will pass through the inaccessible intersection of two given lines.

66. Given a parallelogram $ABCD$ with a circle passing through A, cutting $\overline{AB}$, $\overline{AC}$, $\overline{AD}$ in P, Q, R respectively. Prove: $AB \cdot AP + AD \cdot AR = AC \cdot AQ$.

67. M is the midpoint of chord $\overline{AB}$ of a circle. Any other chord $\overline{CD}$ is drawn through M so that the order of points is A, D, B, C. Tangents drawn at C and D meet $\overleftrightarrow{AB}$ in P and Q respectively. Prove: $CP = QD$, and $PA = QB$.

68. Point E is taken on side $\overline{BA}$ of equilateral triangle ABC, and D is taken on $\overline{AC}$ so that $\frac{BE}{EA} = \frac{AD}{DC} = \frac{1}{2}$. $\overline{BD}$ intersects $\overline{CE}$ at F. Prove that $\overline{AF}$ is perpendicular to $\overline{EC}$.

III

PROBLEMS WITH AN ALGEBRAIC BACKGROUND

69. If Bob can beat Jim by one-tenth of a mile in a two-mile race, and Jim can beat Henry by one-fifth of a mile in a two-mile race, by what distance could Bob beat Henry in a two-mile race?

70. In a race of d yards, **A** beats **B** by p yards, **B** beats **C** by q yards, and **A** beats **C** by r yards. Express d in terms of p, q, r.

71. Prove: If a straight line intersects the graph of $xy=k$ at points A and B and the coordinate axes in points C and D, then $AC=DB$.

72. In a 120-mile boat race we are required to go 60 miles upstream and return to our starting point. We complete the course with all possible speed in a motor boat which is capable of 24 miles per hour in still water. We find that our average speed for the round trip is 18 miles per hour. What is the rate of the current? What data is given in this problem which is not needed for its solution?

73. Towns **A** and **B** are located on a straight river, **A** being 24 miles downstream from **B**. The river is navigable for only half the distance so that a person can row only 12 miles and must walk the other 12. Under these circumstances a man can travel from **B** to **A** in five hours, and from **A** to **B** in seven hours. If there were no current, the journey would require $5\frac{2}{3}$ hours. Find his rate of walking, his rate of rowing, and the rate of the current.

74. Some hikers start on a walk at 3:00 p.m. and return at 9:00 p.m. If their speed is four miles per hour on level land, three miles per hour uphill, and six miles per hour downhill, how far did they walk?

75. Two cows in four weeks can eat all of the grass on two acres and all the grass that grows on these two acres during the four weeks. Three cows in two weeks can eat all the grass on two acres and all the grass that grows on these two acres during the two weeks. How many cows can eat in six weeks all the grass on six acres and all the grass that grows on these six acres during the six weeks? Assume that the rate of growth of the grass is constant and that the grass is evenly distributed.

76. On Christmas Eve two candles, one of which was one inch longer than the other, were lighted. The longer one was lighted at 4:30 and the shorter one at 6:00. At 8:30 they were both the same length. The longer one burned out at 10:30, and the shorter one at 10:00. How long was each candle originally?

77. A commuter always took the 5 o'clock boat and was met by his chauffeur, who always left his home just in time to meet that boat as it docked. One afternoon he caught the 4 o'clock boat and, instead of waiting at the dock, he started up the road. When he saw his car coming, he flagged it, got in, and rode home, arriving a half hour earlier than usual. How long was he walking?

78. For what values of x is $x^2-6x+8>0$?

79. If a and b are the roots of the equation $x^2-px+q=0$, find the equation whose roots are $a+\frac{1}{b}$ and $b+\frac{1}{a}$.

80. A tank contains 250 gallons of water. A certain quantity of water is dipped out and replaced by the same amount of alcohol. From the mixture, the same amount of mixture is removed and is replaced by the same amount of alcohol. The final mixture contains water and alcohol in the ratio 16:9 respectively. How much water was dipped out each time?

81. Two candles of equal length start burning at the same time. One of the candles will burn in 4 hours, and the other in 5 hours. How long will they have to burn before one candle is 3 times the length of the other?

82. Find 3 integers that form a geometric progression if their sum is 21 and the sum of their reciprocals is $\frac{7}{12}$.

83. Solve the system of equations:

$$\frac{y-x}{xy}=\frac{1}{36}$$
$$xy(y-x)=324.$$

84. Solve the system of equations:

$$x^2+y^2+x+y=8$$
$$xy+x+y=5$$

85. If $a=\frac{x}{x^2+y^2}$ and $b=\frac{y}{x^2+y^2}$, solve for x and y in terms of a and b.

86. Find the least number of terms of the geometric progression: $N=1+\frac{1}{2}+\frac{1}{4}+\ldots$ such that $2-N<.0001$.

87. **A**, **B**, and **C** are thermometers with different scales. When **A** reads 12° and 36°, **B** reads 13° and 29°, respectively. When **B** reads 20° and 32°, **C** reads 57° and 84°, respectively. If the temperature drops 18° using **A**'s scale, how many degrees does it drop using **C**'s scale?

88. A certain operation, in the form of an algebraic monomial, was performed on x, producing y. The same operation was then performed on y producing $3x^2$. Find y.

89. You may have been taught that the graphs of two quadratic equations can intersect in no more than four points. How then is it possible for the coordinates of the five points $(-1, 6)$, $(3, -2)$, $(1, 1)$, $(0, 4)$, and $(2, 0)$ to satisfy both of these quadratic equations?

(1) $2x^2-y^2-xy-4x+4y=0$;

(2) $6x^2-y^2+xy-16x+2y+8=0$.

90. A treasure is located at a point along a straight road with towns A, B, C, and D on it in that order. A map gives the following instructions for locating the treasure:

(1) Start at town A and go $\frac{1}{2}$ of the way to C.

(2) Then go $\frac{1}{3}$ of the way to D.

(3) Then go $\frac{1}{4}$ of the way to B, and dig for the treasure.

If $AB=6$ miles, $BC=8$ miles, and the treasure is buried midway between A and D, find the distance from C to D.

91. A second treasure is buried somewhere along another straight road on which are points O, P, Q, R, and S in that order.

The coordinates of towns P, Q, R, and S are x_1, x_2, x_3, x_4 respectively, with respect to origin O. Instructions for finding the treasure read exactly as in Problem 90, except that the names of the towns in the instructions are illegible. Nevertheless, the location of this treasure can be calculated. How?

92. If $x^2+ax+bc=0$ and $x^2+bx+ca=0$ $(a,\ b,\ c\neq 0)$ have a single common root, prove that their other roots satisfy $x^2+cx+ab=0$.

93. If the quadratic equation $x^2+px+q=0$ has two real roots, they can be located graphically as follows. On a set of axes locate $A\ (0,\ 1)$ and $B\left(\frac{-p}{2},\frac{q+1}{2}\right)$. With B as center draw a circle passing through A. The circle will intersect the x-axis in two points whose abscissas are the roots of the given equation. Prove this. What happens if the roots are imaginary?

IV

INFERENTIAL PROBLEMS

94. The organization of the business office of a certain company consists of the following: Manager, Assistant Manager, Cashier, Teller, Treatment Clerk, and Stenographer. The names of the personnel of the office in alphabetical order are: Mr. Brown, Mr. Cowan, Miss Gordon, Mrs. Johnson, Miss Leonard, and Mr. O'Shaunessy.

(a) The Assistant Manager is the Manager's grandson.

(b) The Cashier is the Stenographer's son-in-law.

(c) The Teller is Miss Gordon's step-sister.

(d) Mr. Brown is a bachelor.

(e) Mr. Cowan is 25 years old.

(f) Mr. O'Shaunessey is the Manager's neighbor.

What is the Cashier's name?

95. "Are those your children that I hear playing in the garden?" asked the visitor. "There are really four families of children," replied the host. "Mine is the largest, my brother's family is smaller, my sister's is smaller still, and my cousin's is the smallest of all. They are playing drop the handkerchief," he went on. "They prefer baseball, but there are not enough children to make two teams. Curiously enough," he mused, "the product of the numbers in the four groups is my house number, which you saw when you came in."

"I am something of a mathematician," said the visitor, "let me see whether I can find the numbers of children in the various families." After figuring for a time he said, "I need more information. Does your cousin's family consist of a single child?" The host answered his question, whereupon the visitor said, "Knowing your house number, and knowing the answer to my question, I can now deduce the exact number of children in each family."

How many children were there in each of the four families?

96. Jack Donavan was killed on a lonely road two miles from Trenton at 3:30 a.m. on March 17, 1933. Shorty Malone, Tony Verelli, Hank Rodgers, Joey Freiberg, and Red Johnson were arrested a week later and questioned. Each of these men made four simple statements of which three were absolutely true and only one of them false. One of these five men killed Donavan.

Shorty: "I was in Chicago when Donavan was murdered. I never killed anyone. Red is the guilty man. Joey and I were pals."

Hank: "I did not kill Donavan. I never owned a revolver in my life. Red knows me. I was in Philadelphia the night of March 17."

Tony: "Hank lied when he said he never owned a revolver. The murder was committed on St. Patrick's Day. Shorty was in Chicago at that time when the murder was committed. One of us five is guilty."

Joey: "I did not kill Donavan. Red has never been in Trenton. I never saw Shorty before. Hank was in Philadelphia with me the night of March 17."

Red: "I did not kill Donavan. I have never been in Trenton. I never saw Hank before now. Shorty lied when he said I'm guilty."

Which one of the five men killed Donavan?

97. A motorist is driving in the state of Euphoria when he comes to a fork in the road. He knows that one of these roads leads to Avon but he does not know which. Standing at the fork of the road is a man who may or may not be a Euphorian. The motorist has been informed that Euphorians always lie, that non-Euphorians always tell the truth, and that this individual knows which road leads to Avon. The motorist is allowed to ask only one question. What question can he ask which will enable him to determine which road leads to Avon?

98. Thalia Collingwood's father has a yacht and so has each of his four friends, Colonel Drake, Mr. Nelson, Sir Barnacle Hood, and Dr. Benbow. Moreover, each of the five has just one daughter, and each has named his yacht after a daughter of one of the others.

So there is a yacht named after Thalia. Sir Barnacle's yacht is the *Gabrielle;* Mr. Collingwood owns the *Lorna;* Mr. Nelson the *Rosalind.* The *Melissa,* owned by Colonel Drake, is named after Sir Barnacle's daughter. Gabrielle's father owns the yacht which is named after Dr. Benbow's daughter. Who is Lorna's father?

99. A sloop, steamer, and oil tanker are named the *Washington,* the *Lincoln,* and the *Jefferson,* but not necessarily in that order. Their ports of departure and destination are Bermuda, Boston, Halifax, London, Newport News, and New York, here listed alphabetically.

(1) The oil tanker passed the ship that was going to Bermuda.

(2) The *Lincoln* arrived in Halifax the same day that the steamship left London.

(3) The *Washington* left from New York under full sail and did not go to Boston, although one of the others did. What were the names of the sloop, the steamship, and the oil tanker, and what were their respective ports of departure and destination?

100. John, Karl, Luke, Mark, and Oswald are freshman, sophomore, junior, senior, and graduate students in dentistry, engineering, medicine, law, and pharmacy, but not necessarily respectively.

(1) Mark is in a higher class than Karl.

(2) John will become an intern next year.

(3) The engineering student was graduated from the same high school as Karl and Luke.

(4) Oswald was graduated from the law school last year with honors.

(5) Karl and the sophomore pharmacy student room together.

Pair off the boys according to profession and class in school.

101. Brown, Todd, Robinson, Slater, Mitchell, O'Brien, Grant, and Ford reached the final round of the amateur boxing championships. They were the finalists in the 118-pound, 126-pound, 145-pound, and heavyweight classes.

(1) Brown weighed in at 117 pounds, and Todd was the heaviest winner of all.

(2) Grant defeated Mitchell, and Slater won by a knockout.

(3) O'Brien lost the 145-pound final by a knockout, and Ford defaulted in the bantamweight final.

Who was the champion and who was the runner-up in each division?

102. One of five brothers had broken a window. John said, "It was Henry or Thomas." Henry said, "Neither Ernest nor I did it." Thomas said, " You are both lying." David said, "No, one of them is speaking the truth, but not the other." Ernest said, "No, David, that is not true." When the father of the boys (a truthful man), was consulted, he said, "Three of my sons always speak the truth, but I am afraid you cannot always rely on what the other two say." Who broke the window?

103. (1) Joe, Jack, John, and Tom live one on each floor of a four-story apartment.

(2) Their ages are 10, 9, 8, 5, but not necessarily in that order.

(3) Joe lives directly above the 9-year-old and directly below the 8-year-old.

(4) Jack has to pass by the 5-year-old to leave the building from his apartment.

(5) Jack is more than one floor away from Tom, who is more than one year younger than Jack.

Find the ages and on which floor each of the boys lives.

104. Seven cross-country runners, each bearing a number on his back from 1 to 7 (no duplicates) cross the tape at the end of a race, the first at 12:01, and the others at intervals of one minute apart, with no ties.

From the following facts determine the order in which the runners came in. In each statement "time" means the number of minutes after 12:00.

(1) The highest total of runner and his time, viz. 13, occurs just once.

(2) The lowest total of runner and his time, that is 4, occurs just once.

(3) The total 9, of runner and time, occurs just three times.

(4) The winner's number exceeds that of the second to come in.

105. Restore the missing numerals in this multiplication:

$$
\begin{array}{r}
4\,*\,* \\
3\,* \\
\hline
3\,6\,*\,* \\
\,\,7\,* \\
\hline
\,\,3\,*\,*
\end{array}
$$

106. Each of the letters in the following addition problem represents a different digit. Reconstruct the problem in numerals:

$$\begin{array}{r} A\,H\,A\,H\,A \\ T\,E\,H\,E \\ \hline T\,E\,H\,A\,W \end{array}$$

107. Each of the letters in the following division problem represents a different digit. Reconstruct the problem in numerals:

```
          R I R
      ___________
  G R ) L M T T R
        L T R
        -----
          V V T
          V L S
          -----
            L T R
            L T R
            -----
```

108. The product of a four-digit number *abcd*, $(a \neq 0)$ by 1999 is a number ending with four repeated digits. Find all possible numbers *abcd*.

109. If the repeated digit in the answer to Problem 108 is t, express a, b, c, and d in terms of t.

V

MISCELLANEOUS PROBLEMS

110. Of the members of three athletic teams in a certain school, 21 are on the basketball team, 26 on the baseball team, and 29 on the football team. 14 play baseball and basketball; 15 play baseball and football, and 12 play football and basketball. 8 are on all three teams. How many members are there altogether?

111. Find the least two positive integers having the remainders 2, 3, 2 when divided by 3, 5, 7, respectively.

112. Prove that $2^n \cdot 3^{2n} - 1$ is divisible by 17 for all positive integral values of n.

113. A dollar and sixty-two cents is made up from pennies, dimes, and nickels; half of the coins are nickels. Find the number of ways in which this division can be done, giving the number of pennies, dimes, and nickels for each solution.

114. The positive integers are arranged in groups as follows: (1), (2, 3), (4, 5, 6), (7, 8, 9, 10), ... Note that there is one number in the first group, there are two numbers in the second, and so on. Find and prove a formula for the sum of the k numbers in the kth group.

115. In a certain classroom, the 25 seats are arranged in a square of five rows, with five seats in each row. The teacher announces that each student is to change his seat by going either to the seat in front of him, in back of him, or to the right or left of him. Can this order be carried out? Prove your answer.

116. In each game make the black and white pieces interchange their positions by moving as follows:

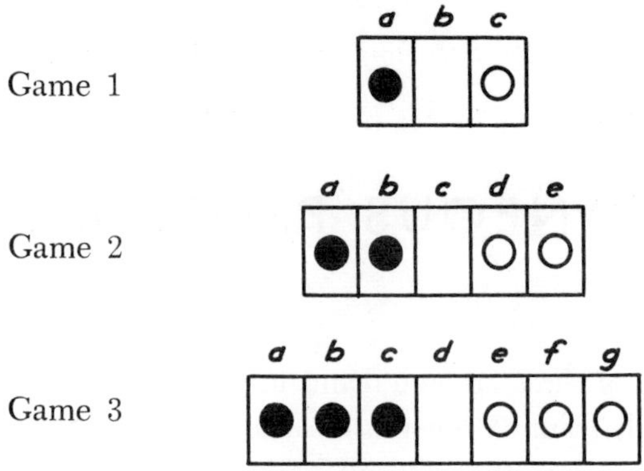

Figure 10

(a) Black may move to the right only; white to the left only, each into an empty space.

(b) Each may jump over a single opponent provided it then lands on an empty space. All pieces remain on the board; there are no captures as in checkers.

Questions

(1) Make the pieces change places in each of these games.

(2) What is the minimum number of moves that it should take for n black pieces and n white pieces on a board with $2n+1$ squares?

(3) What is the minimum number of moves that it should take for b black pieces and w white pieces on a board with $b+w+1$ squares?

117. In a new variation of the game of NIM, two players **A** and **B** start with a pile of N matches and alternate in removing them. **A**, playing first, removes any number of matches, from one to not more than half the pile. **B** then removes any number from one to not more than half the remaining pile. Alternating turns (with the additional rule that if a single match remains it may be removed), the one who removes the last match is the winner.

(a) Show that, with best play on both sides, **A** will win when $N=1, 3, 4,$ or 6; **B** will win when $N=2$ or 5.

(b) If N_x is the xth value for which **B** will win, express N_{x+1} in terms of x.

118. The average of the six permutations of 1, 2, 8, viz. 128, 182, 218, 281, 812, 821 is $\frac{2442}{6}$ or 407. It is observed that **(a)** the average is an integer; **(b)** $1+2+8=4+0+7$. Are these observations valid for any three digits?

119. Examine the following numerical relationships:

$$1^3=1^2-0^2$$

$$2^3=3^2-1^2$$

$$3^3=6^2-3^2$$

$$4^3=10^2-6^2$$

What generalization do they suggest? Prove it.

120. A man and his grandson have the same birthday. For six consecutive birthdays the man is an integral number of times as old as his grandson. How old is each at the sixth of these birthdays?

121. A certain light signal goes on at precisely 12 noon. Thereafter it goes off and on at equal intervals, each lasting an integral number of minutes. It is observed that at 12:09 the light is off; at 12:17 it is on; and at 12:58 it is on. Will the light be on or off at two o'clock?

122. On a Cartesian set of axes we call any point with integral coordinates a lattice point.

A rectangle has vertices at lattice points, and its sides are parallel to the axes. Let I represent the number of interior lattice points of of the rectangle; S the number of lattice points on its sides; and A the area of the rectangle. Express A in terms of I and S.

123. **(a)** Find all exact divisions in which one subtracts first 1296, then 432, then 864;

(b) Find the dividend of an exact division in which one subtracts first A, then B, then C.

124. Examine the following numerical relations:

(1) 91 and $9-(2\times 1)$, (or 7) are each divisible by 7;

(2) 385 and $38-(2\times 5)$, (or 28) are each divisible by 7;

(3) 4816 and $481-(2\times 6)$, (or 469) are each divisible by 7.

What generalization do they suggest? Prove it.

125. In Fig. 11:

(a) Find the sum of the integers in the cells of each diagonal (e.g. $1+6+11+16$). Do the same for similarly numbered squares 2 by 2, 3 by 3, 5 by 5, etc.

(b) What generalization do the results suggest? Prove it.

1	2	3	4
5	6	7	8
9	10	11	12
13	14	15	16

Figure 11

126. In the series $+3-2+3-2+\ldots$, in which $+3$'s and -2's alternate, find a formula for the nth term.

127. Prove that it is impossible at any time for the hour, minute, and second hands of a clock—all attached at the center of its face—to divide the face into three equal sectors.

128. In a 10-team baseball league, each team plays each of the others 18 times. The season ends, not in a tie, with each team the same number of games ahead of the following team. What is the greatest number of games that the last team could have won?

129. Prove that a positive integer greater than nine, all of whose digits are alike, cannot be a perfect square.

130. Place a ruler and a cylindrical object, such as a spool or pencil, on a flat, level surface, as in Fig. 12. Move the ruler to the left

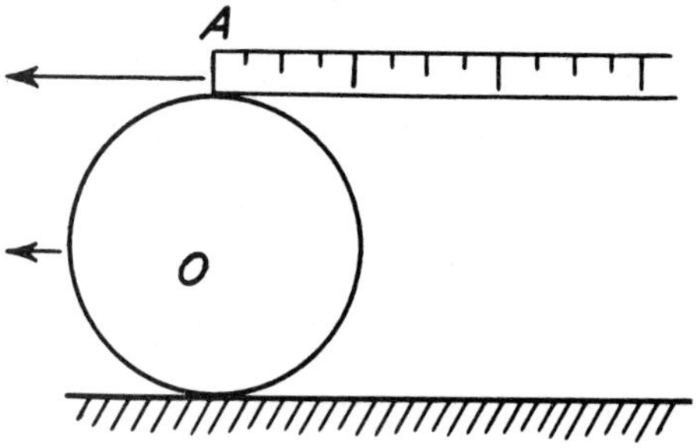

Figure 12

keeping it parallel to the surface, thus rolling the cylinder also to the left. Compare the distances that A on the ruler and center O of the circle move. Do this for various distances and radii. What formula relating these distances is suggested by your data? Explain how you might have predicted this on mathematical grounds.

131. The pages of a book are numbered consecutively from 1 to n. They are divided into three sections such that the number of digits used to number the pages in each of the sets of pages is the same. The three smallest values of n for which this is possible are 3, 6, 9. What is the next smallest value of n, and at what pages do the sections end?

132. If n is a positive integer, express in terms of n the number of solutions in positive integers of $2x+y+z=n$ when **(a)** n is even; **(b)** n is odd.

133. **(a)** Show that three coins in a row, all heads, can be turned to all tails under the following conditions: (1) only one coin at a time may be turned; (2) each of the eight possible heads-tails arrangements shall appear once and only once, ending with **TTT**. For example: (1) **HHH**; (2) **HHT**; (3) **HTT**; (4) not **TTT**, which must occur at step 8.

(b) Prove that there can be no solution for four coins, under the same conditions. (There are now 16 possible heads-tails arrangements.)

134. A greeting card dealer has three kinds of packets. Packet **A** contains 3 get-well cards, 1 birthday card, and 7 anniversary cards. Packet **B** contains 2, 3, and 1 respectively. Packet **C** contains 5, 4, and 2, respectively. He wishes to merge these so as to form new packets, each containing *one* get-well card, *one* birthday card, and *one* anniversary card.

What is the least number of packets **A**, **B**, and **C** that must be used to accomplish this? How many new packets will be made?

135. Prove without the use of tables that $(\sin 12°)(\sin 48°)(\sin 54°)=\frac{1}{8}$.

136. Prove that in any triangle ABC, $\cot A+\cot B+\cot C=\dfrac{a^2+b^2+c^2}{\triangle}$, where $\triangle$ represents the area of triangle ABC.

137. A dealer bought some radios for \$432, paying a dollars for each. He sold all but 18 of them for \$324, receiving b dollars for each. If a and b are integers, and if he gained on the sale of each radio, how many radios did he buy originally?

138. Find all the integers between 200 and 300 that have exactly six positive factors.

139. Enclosures such as those in Fig. 13 are constructed of stair-patterned sets of squares. If $n=$ the number of squares on a side, and $T=$the number of walls, then by actual count, for $n=1, 2, 3, 4$, $T=4, 10, 18, 28$.

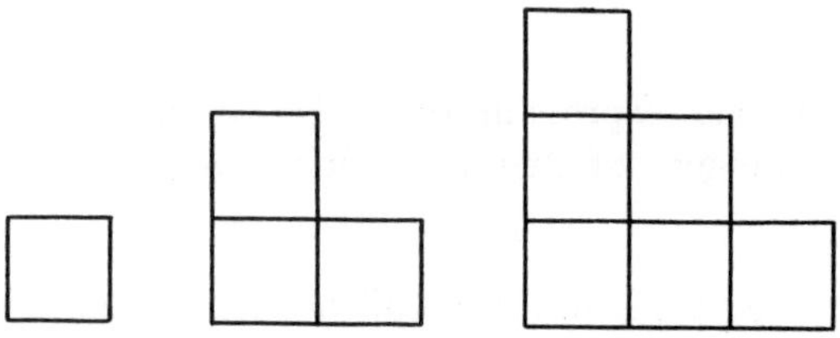

Figure 13

317 walls are available for constructing such an enclosure, but are not sufficient to complete it. What is the least number of additional walls needed to complete the stair-pattern?

140. In the product $9 \cdot HATBOX = 4 \cdot BOXHAT$, find the six-digit numbers, $HATBOX$ and $BOXHAT$.

SOLUTIONS TO PROBLEMS

I

A MISCELLANY OF BREVITIES

1. Striking six times creates five intervals in five seconds, or one interval per second. In ten seconds ten intervals must be created. This will require striking eleven times.

A related problem is that of determining the number of parallel lines needed in the theorem: *If three or more parallel lines cut off equal segments on one transversal, they cut off equal segments on any other transversal.* For n equal segments, $n+1$ parallel lines are needed.

In a variation of the above idea, one is asked for the cost of cutting a log into eight equal segments, if cutting it into four equal segments costs 60 cents. This is equivalent to asking if three cuts cost 60 cents, how much will seven cuts cost?

2. **(a)** The Unique Factorization Theorem for the realm of integers assures that 180 can be expressed uniquely as the product of powers of primes.
Reference: [16].

(b) A perfect square must contain each prime factor an even number of times.

(c) A perfect cube must contain each prime factor a multiple-of-three number of times.

By (**a**) $180=2^2\cdot3^2\cdot5$. By (**b**) another 5 is needed to make the product a perfect square. By (**c**) $2\cdot3\cdot5^2$ is needed to make the product a perfect cube.

Note:

Unique factorization is not valid for all number realms. For example, in the realm of *algebraic numbers*, $21=3\cdot7=(4+\sqrt{-5})(4-\sqrt{-5})=(1+2\sqrt{-5})(1-2\sqrt{-5})$.

An algebraic number is one which is the root of some algebraic equation, i.e. one of the form:

$$a_n x^n + a_{n-1}x^{n-1} + \cdots + a_1 x + a_0 = 0, \ (n \geqq 1, a_n \neq 0),$$

where the n's are integers. For example, 3 is the root of $x-3=0$; $4-\sqrt{-5}$ is a root of $x^2-8x+21=0$; $1+2\sqrt{-5}$ is a root of $x^2-2x+21=0$.

3. (a) Pages 1-9 use 9 digits; pages 10-99 use 180 digits. This leaves 663 digits to make the required total of 852 digits. These are obtained by going 221 pages beyond 99, viz. page 320.

   (b) The number of 7's used in pages 1 to 100, 101 to 200, and 201 to 300 are 20 each. This number added to the two 7's in 307 and 317 makes a total of 62 7's.

4. Since his total expenditure is \$140, and his total income is \$160, his profit is \$20.

5. $\frac{30}{7}$.

6. It would not matter, because all six possible orders would yield the same final result.

   After a discount of a percent from list price L, there remains $\left(1-\frac{a}{100}\right)L$. If this is followed by a discount of b percent, there remains $\left(1-\frac{b}{100}\right)$ times the previous amount, or $\left(1-\frac{b}{100}\right)\cdot\left(1-\frac{a}{100}\right)L$. After a third discount of c percent, there remains $\left(1-\frac{c}{100}\right)\cdot\left(1-\frac{b}{100}\right)\cdot\left(1-\frac{a}{100}\right)L$. This result is obviously not affected by the order of a, b, c.

   After discounts of 20 percent, 10 percent, 5 percent (in any order) there remains $(1-.20)(1-.10)(1-.05)$, or 68.4 percent of the list

7. If S is the man's salary and x is the required percent, $S\left(1-\frac{p}{100}\right)\left(1+\frac{x}{100}\right)=S$. Hence $x=\frac{100p}{100-p}$.

8. Let s, p, m represent selling price, profit, and marked price. Then $s=c+p$, and $s=m\left(1-\frac{d}{100}\right)$. Also $p=\frac{ds}{100}=\left(\frac{dm}{100}\right)\left(1-\frac{d}{100}\right)$. When we substitute these values of s and p in the first equation and solve for m, we obtain: $m=\frac{c}{\left(1-\frac{d}{100}\right)^2}$

9. **(a)** By definition: **Average Rate** $=\dfrac{\textbf{Total Distance}}{\textbf{Total Time}}$. If the distance traveled from home to work is d miles, the time taken was $\dfrac{d}{15}$. Similarly, the time taken to return the same distance was $\dfrac{d}{30}$. Hence Average Rate $=\dfrac{2d}{\dfrac{d}{15}+\dfrac{d}{30}}=20$ miles per hour, a result which is independent of the distance.

At first sight this may appear wrong, because surely the average of 15 and 30 is one half their sum or $22\frac{1}{2}$. But consider that at 30 miles per hour, the vehicle spends less time traveling d miles than at 15 miles per hour. We should not, therefore, take the arithmetic average of two rates that endure for different times.

(b) **Average Rate** $(R)=\dfrac{2d}{\dfrac{d}{x}+\dfrac{d}{y}}=\dfrac{2xy}{x+y}$.

This may be written: $\dfrac{1}{R}=\dfrac{1}{2}\left(\dfrac{1}{x}+\dfrac{1}{y}\right)$. Thus the *reciprocal* of the average rate is the arithmetic average of the *reciprocals* of the two rates. R is called the Harmonic Mean between x and y.

10. Two formulas are: $(-1)^{n+1}$ and $\cos(n+1)\pi$. Others can be constructed. See Problem 126.

11. The new root y is to be the reciprocal of the old root x. Therefore let $y=\dfrac{1}{x}$, or $x=\dfrac{1}{y}$. Substituting in **(a)** and **(b)** there result: **(a)** $4x^2-11x+6=0$; **(b)** $cx^2+bx+a=0$.

Note:

This is a special case of the more general problem of transforming a given equation, $f(x)=0$, into one whose roots are each a certain function of the roots of $f(x)=0$. The purpose of such a transformation could be to obtain an equation that is more convenient to work with. For example, to solve $ax^2+bx+c=0$, let $y=x+\dfrac{b}{2a}$, or $x=y-\dfrac{b}{2a}$. Then $a\left(y-\dfrac{b}{2a}\right)^2+b\left(y-\dfrac{b}{2a}\right)+c=0$. This reduces to $4a^2y^2-b^2+4ac=0$, which is easily solved because the first degree term in y is missing. We have

$y=\frac{\pm\sqrt{b^2-4ac}}{2a}$; and replacing y by $x+\frac{b}{2a}$, we obtain the familiar formula for the roots of a quadratic equation.

Among other transformations that have been found useful are the following:

(i) Increase or diminish the roots by a given constant a: let $y=x+a$; $y=x-a$.

(ii) Multiply or divide the roots by a given constant a: let $y=ax$; $y=\frac{x}{a}$.

(iii) Change the signs of the roots: let $y=-x$.

(iv) Combinations of the above.

Problem

Show that the roots of $ax^4+bx^3+cx^2+bx+a=0$ occur in pairs that are reciprocals of each other. (Let $y=\frac{1}{x}$, and compare the new equation with the given equation.)

12. **(a)** If $x=b^{\log a}$, take the logarithm of both sides to the base b. Thus $\log_b x=(\log_b a)(\log_b b)=\log_b a$; and $x=a$.

A direct application of the definition of a logarithm to a given base yields the same answer at once.

(b) Since $b^{-\log a}$ is the reciprocal of $b^{\log a}$, the answer is $\frac{1}{a}$.

13. Let $r=4t+u$, $(0\leqq u<4)$, $s=4v+w$, $(0\leqq w<4)$. If $i^r=i^s$, then $(i^4)^t\cdot i^u=(i^4)^v\cdot i^w$, or $i^u=i^w$. Therefore r and s each give the same remainder when divided by 4. This condition can be shown to be both necessary and sufficient.

In the language of congruences: $i^r=i^s$ if and only if $r\equiv s \pmod 4$. For the meaning of "$\equiv$," see [16], Chapter 4.

14. Using the result of 13, we obtain: $i^9+i^{10}+i^{11}+i^{12}=i+i^2+i^3+i^4$. Since this is equal to $i+(-1)+(-i)+1$, or zero, the answer to the given problem is *zero*.

In general, the result in Problem 13 shows that the sum of any four consecutive powers of i will be some cyclic arrangement of $i+(-1)+(-i)+1$. Therefore any power of such a sum will be equal to zero.

15. Since $8^{\frac{1}{6}}=(2^3)^{\frac{1}{6}}=2^{\frac{1}{2}}$, and $\frac{7}{3-\sqrt{2}}=3+\sqrt{2}$, (by rationalizing the denominator), the given equation results in $x^{\frac{1}{3}}=3$, or $x=27$.

16. **(a)** If $N=x^2$, $(x+1)^2=x^2+2x+1=N+2\sqrt{N}+1$.

(b) If $N=x^3$, $(x+1)^3=x^3+3x^2+3x+1=N+3\sqrt[3]{N^2}+3\sqrt[3]{N}+1$.

17.

First Solution

Clear fractions. $\sqrt{x+1}+\sqrt{x-1}=3\sqrt{x+1}-3\sqrt{x-1}$. Therefore $4\sqrt{x-1}=2\sqrt{x+1}$, or $2\sqrt{x-1}=\sqrt{x+1}$. Squaring both sides: $4(x-1)=x+1$, and $x=\frac{5}{3}$.

Second Solution

Apply the theorem, "If $\frac{a}{b}=\frac{c}{d}$, then $\frac{a+b}{a-b}=\frac{c+d}{c-d}$." (Appendix A)

Let $a=\sqrt{x+1}$, $b=\sqrt{x-1}$, $c=3$, $d=1$; then $\frac{2\sqrt{x+1}}{2\sqrt{x-1}}=\frac{3+1}{3-1}$, or $\frac{\sqrt{x+1}}{\sqrt{x-1}}=\frac{2}{1}$. Square both sides and apply the same theorem to obtain the answer.

Third Solution

Rationalize the denominator of the left member and solve the resulting radical equation.

18. We have:

$$a+b=17$$
$$b+c=23$$
$$c+a=12$$

Adding the three equations: $2(a+b+c)=52$;
therefore: $a+b+c=26$.

Subtracting each of the first three equations in turn from the last one yields: $c=9$; $a=3$; $b=14$.

19.

$$m\angle A+m\angle 1+m\angle D=180^\circ$$
$$m\angle B+m\angle 2+m\angle E=180^\circ$$
$$m\angle C+m\angle 3+m\angle A=180^\circ$$
$$m\angle D+m\angle 4+m\angle B=180^\circ$$
$$m\angle E+m\angle 5+m\angle C=180^\circ$$

$2(m\angle A+m\angle B+m\angle C+m\angle D+m\angle E)+3\cdot 180^\circ=5\cdot 180^\circ$.
Therefore $m\angle A+m\angle B+m\angle C+m\angle D+m\angle E=180^\circ$.

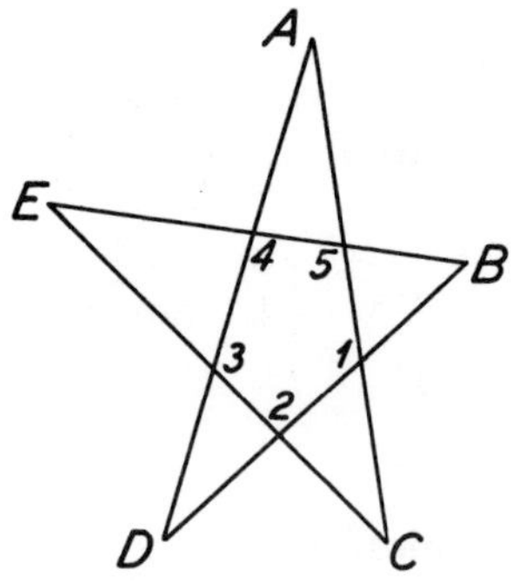

Figure 14

20. Draw $\overleftrightarrow{BG} \| \overleftrightarrow{AD}$ intersecting $\overline{EF}$ at H, as in Fig. 15.

$EF = EH + HF$. But $EH = AB$, and $\frac{HF}{GC} = \frac{BF}{BC} = \frac{2}{3}$. Hence $EF = AB + \frac{2}{3}GC = AB + \frac{2}{3}(DC - AB) = 7 + \frac{2}{3}(10 - 7) = 9$.

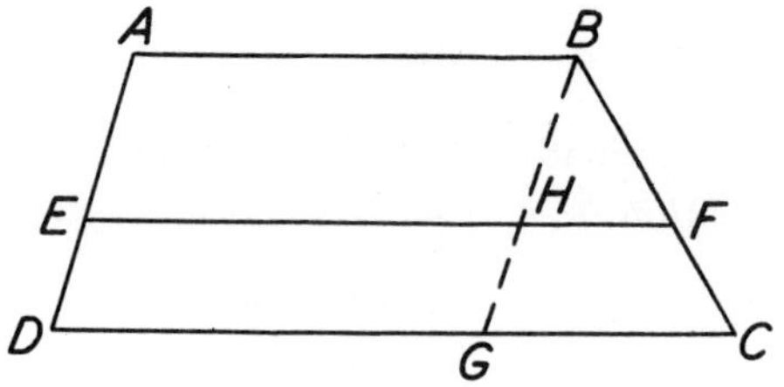

Figure 15

Notes:

(i) The formula $EF = AB + \frac{2}{3}(DC - AB)$ may also be written: $EF = \frac{1 \cdot AB + 2 \cdot DC}{3}$.

(ii) In general, if $\frac{AE}{ED} = \frac{BF}{FC} = \frac{m}{n}$, then by a method similar to that above, it can be proven that $EF = \frac{n \cdot AB + m \cdot DC}{m + n}$.

(iii) The formula in (**ii**) may be applied to determine the coordinates of a point that divides a line segment in a given ratio, if the coordinates of the endpoints are given.

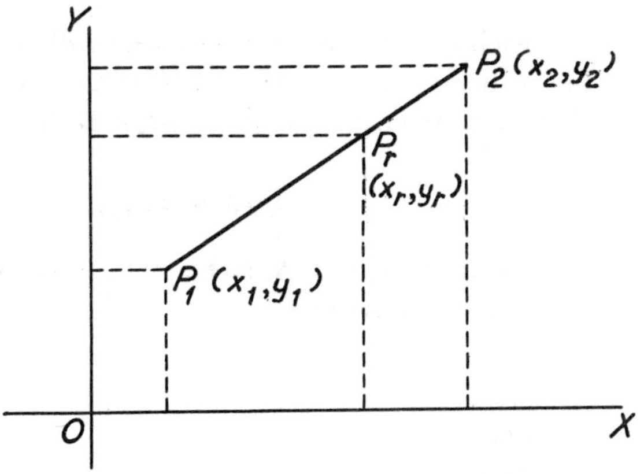

Figure 16

If $\frac{P_1P_r}{P_rP_2}=\frac{m}{n}$, then

$$x_r=\frac{n\cdot x_1+m\cdot x_2}{m+n}$$

$$y_r=\frac{n\cdot y_1+m\cdot y_2}{m+n}$$

These are known as Joachimsthal's formulas.

21. If each side of the base of the box is x, with a height of s, the volume is $x^2s=c$. Therefore $x=\sqrt{\frac{c}{s}}$, and the side of the original square is $x+2s$, or $\sqrt{\frac{c}{s}}+2s$.

22. **(a)** 9^{9-9}.

(b) $.\dot{9}$, which represents the infinite repeating decimal: .999 . . . .

23. Because the left member is divisible by three and the right member is not, there are no integers that satisfy this equation.

24. **First Solution**

The first man shakes hands with seven other men. Then the second man shakes hands with six other men (his handshake with the first man has already been counted). Continuing this way, the number of handshakes comes to: $7+6+5+4+3+2+1=28$.

Second Solution

If each of eight men shakes hands with each of the other seven, the 56 handshakes would consist of 28 sets of duplications, for **A** shaking hands with **B** is the same as **B** shaking hands with **A**.

Third Solution

If the eight men are **A, B, C, D, E, F, G, H,** the handshakes may be enumerated as follows:

AB
AC BC
AD BD CD
AE BE CE DE
AF BF CF DF EF
AG BG CG DG EG FG
AH BH CH DH EH FH GH.

Examination of this table will confirm the methods given above.

25. **(a)** $5+5+5=15$

(b) $999+1=1000$

(c) $89+9=98$

(d) $23+23+23+23=92$.

26. The problem is equivalent to: $\frac{34{,}895}{35}=997$.

27. Assume it is 5 o'clock.
Bill's watch reads 5:10; so he thinks it is 5:15.
John's watch reads 4:50; so he thinks it is 4:40.
Joe's watch reads 4:55; so he thinks it is 4:45.
Henry's watch reads 5:05; so he thinks it is 5:15.
Therefore John and Joe will miss the train.

28. This solution is given in the language of SETS.

It is obvious that $g=7$, $h=12$. The set $\{a, d\}$ must equal the set $\{14, 15\}$, and $\{a, j\}=\{6, 14\}$.

Since $\{a\}=\{a, j\}\cap\{a, d\}$ and $\{14\}=\{14, 15\}\cap\{6, 14\}$, it follows that $a=14$. Hence $d=15$, $j=6$, and $i=13$. Now $\{b, c\}=\{8, 11\}$ and $\{b, e\}=\{4, 8\}$. Since $\{b\}=\{b, c\}\cap\{b, e\}$ and $\{8\}=\{8, 11\}\cap\{4, 8\}$, we have $b=8$. So $c=11$, $e=4$, and $f=10$.

Figure 17

a 14	1	b 8	c 11
d 15	5	e 4	f 10
2	16	9	g 7
3	h 12	i 13	j 6

Note:

As the reader probably knows, this is called a Magic Square. In a square n by n, using consecutive integers from 1 to n^2, one should expect the sum in any column, row, or diagonal to be $\frac{1+2+\ldots+n^2}{n}=\frac{n^2}{2}\cdot\frac{1+n^2}{n}=\frac{n(n^2+1)}{2}$. For $n=4$, this gives the magic sum 34.

II

PROBLEMS WITH A GEOMETRIC BACKGROUND

29. The two lanes have semicircular ends which together form circles with circumferences c_1 and c_2. The difference between the lengths of the two lanes is $d = c_2 - c_1$. If r is the radius of the c_1 circle, in yards, then $r+1$ is the radius of the c_2 circle. Hence $d = 2\pi(r+1) - 2\pi r = 2\pi$. The solution is thus independent of y.

Notes:

(i) A similar problem with a surprising answer is the following:

A belt is stretched tight about the equator of the earth. How long a piece should be inserted in the belt so that it can circle the earth 1 foot away from the equator at all points?

By the same method as that of the above problem it will be found that the extra length of belt should be a mere 2π feet.

(ii) In the above problem, most people would expect a considerably larger answer. In contrast, consider the following problem.

An unbroken trolley track two miles long expands uniformly a length of two feet, causing the track to buckle in the middle and rise so as to form an isosceles triangle. How high will the vertex of the triangle then be above the ground?

Solution

The require distance d is one leg of a right triangle whose other leg and hypotenuse are 5280 feet and 5281 feet long, respectively. Then $d^2 = (5281)^2 - (5280)^2 = (5281 + 5280) \cdot (5281 - 5280)$. Hence $d = \sqrt{10{,}561}$, which gives a result over 100 feet long.

30. Because $\triangle DEF \sim \triangle DAB$, $\frac{EF}{AB} = \frac{DF}{DB}$. And because $\triangle BEF \sim \triangle BCD$, $\frac{EF}{CD} = \frac{BF}{DB}$. By addition we obtain: $\frac{DF+BF}{DB} = \frac{EF}{AB} + \frac{EF}{CD}$. Divide through by EF and replace the left member by its equal 1, to obtain the required result.

Note:

Thus EF is twice the harmonic mean between AB and CD. (See the solution to Problem 9 for the meaning of "harmonic mean.")

31. $\overline{AI}$ and $\overline{BI}$ are angle bisectors. Therefore if the measures of $\angle A$ and $\angle B$ are x and y respectively, $m\angle BAI = \frac{x}{2}$, $m\angle ABI = \frac{y}{2}$, and the measure of exterior angle BID of $\triangle ABI = \frac{x}{2} + \frac{y}{2}$. Now $\angle CBD$ intercepts arc DC and therefore has the same measure as $\angle DAC$, viz. $\frac{x}{2}$. Hence $m\angle DBI = \frac{x}{2} + \frac{y}{2} = m\angle BID$. The rest follows easily.

32. If s is the length of each side of rhombus $ABCD$ (with $\angle A$ acute), and d_1, d_2 are the lengths of its diagonals, then the area of the rhombus may be written in two ways: $s^2 \sin A$, and $\frac{d_1 \cdot d_2}{2}$. But

$\frac{d_1}{s} = \frac{s}{d_2}$; therefore $s^2 = d_1 \cdot d_2$. Thus $\sin A = \frac{1}{2}$, and $m\angle A = 30°$.

33. **(a)** Let a, b, x represent the lengths of $\overline{BC}$, $\overline{AC}$, and the side of inscribed square $EGCF$. Since $\frac{GE}{CB} = \frac{AG}{AC}$, $\frac{x}{a} = \frac{b-x}{b}$, and $x = \frac{ab}{a+b}$.

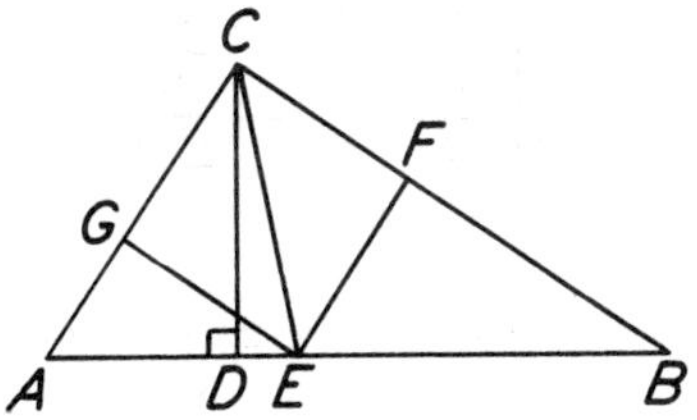

Figure 18

Now twice the area $\triangle ABC = ab = ch$; and $a+b=\sqrt{(a+b)^2}$. Therefore $a+b=\sqrt{a^2+2ab+b^2}=\sqrt{c^2+2ch}$. Hence $x=\dfrac{ch}{\sqrt{c^2+2ch}}$.

(b) **First Method**

Let a, b, y represent the lengths of BC, AC, and the sides of inscribed square $HIJK$. Since $\frac{IJ}{AB}=\frac{CL}{CD}$, $\frac{y}{c}=\frac{h-y}{h}$, and $y=\frac{ch}{c+h}$.

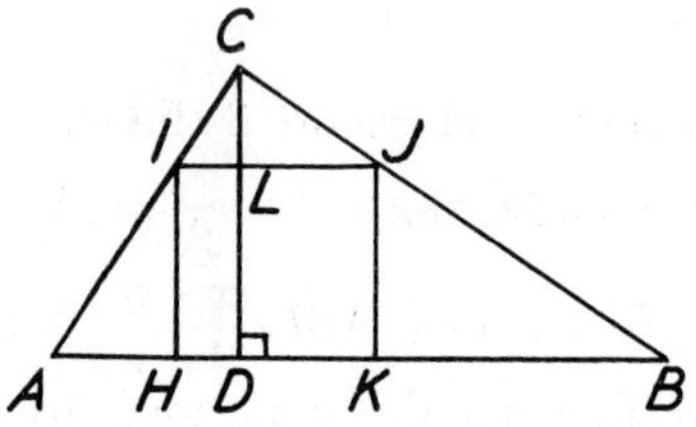

Figure 19

Second Method

To inscribe square $HIJK$ in $\triangle ABC$, proceed as follows: Construct a square $CDEF$ on $\overline{CD}$ as side. Draw $\overline{AF}$ intersecting $\overline{CB}$ in J. Construct rectangle $HIJK$ as in Fig. 20. Since $\frac{HI}{CD}=\frac{AI}{AC}=\frac{IJ}{CF}$, and $CD=CF$, it follows that $HI=IJ$, and $HIJK$ is a square.

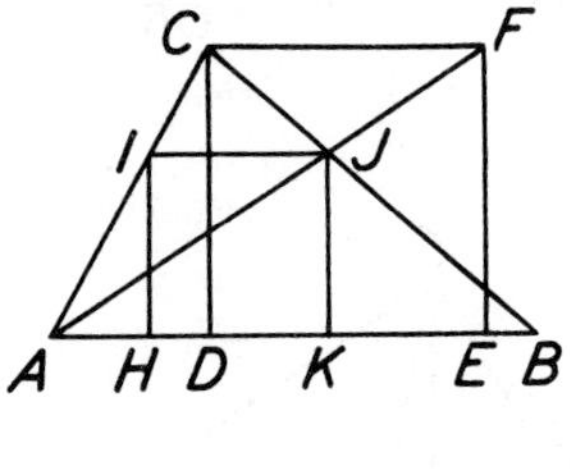

Figure 20

Now consider Fig. 21 extracted from the one above. From the theorem in Problem 30, it follows that $\frac{1}{IJ}=\frac{1}{CF}+\frac{1}{AB}$. Thus

$\frac{1}{y}=\frac{1}{h}+\frac{1}{c}$; that is, the side of the inscribed square is $\frac{1}{2}$ the harmonic mean (see solution to Problem 9) between the hypotenuse and the altitude to the hypotenuse.

Figure 21

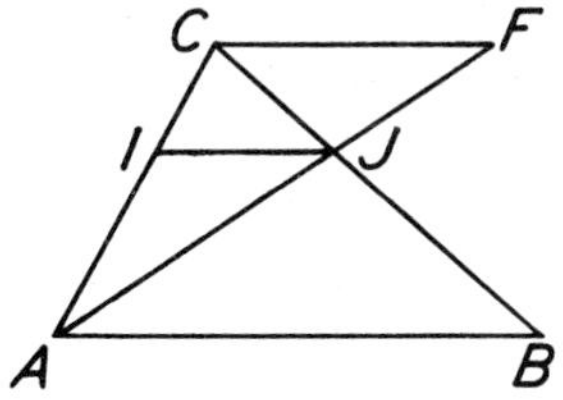

(c) Since $x=\frac{ch}{\sqrt{c^2+2ch}}>\frac{ch}{\sqrt{c^2+2ch+h^2}}=y$, it follows that $x>y$.

34. Let the altitude of the upper and lower trapezoids have lengths m and n respectively. Since the areas are equal, $\frac{m(a+x)}{2}=\frac{n(b+x)}{2}$ or $\frac{m}{n}=\frac{b+x}{a+x}$. Now $\frac{BN}{NC}=\frac{m}{n}$. Therefore, by the second formula in the solution of Problem 20, $x=\frac{na+mb}{m+n}$. Divide numerator and denominator of the right member by n: $x=\frac{a+\frac{mb}{n}}{\frac{m}{n}+1}$. Now substitute $\frac{m}{n}=\frac{b+x}{a+x}$ and solve for x to obtain: $x=\sqrt{\frac{a^2+b^2}{2}}$.

Figure 22

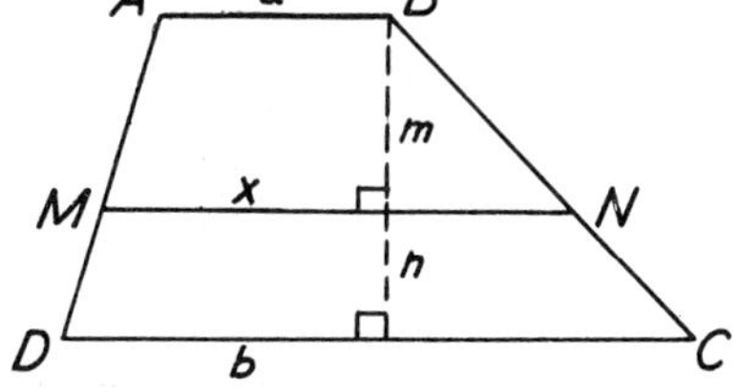

35. Let the given points be A and B, and the given line t. Obviously A and B must lie on the same side of t. There are two possibilities: **(a)** $\overline{AB}$ intersects t at P (say); **(b)** $\overline{AB}||t$.

Case a

Let $\overline{PT}$ be the tangent to the desired circle. Then $\frac{PB}{PT}=\frac{PT}{PA}$. Find PT by the construction for a mean proportional, and use this to locate T. This leads to two solutions, since T can occur on either side of P. The center of the required circle is the intersection of the perpendicular bisector of $\overline{AB}$ and the perpendicular to t at T.

Case b

Draw the perpendicular bisector of $\overline{AB}$ and let it intersect t at T. This is obviously the point of tangency of the required circle. The circle can now be constructed through the three points A, B, and T.

36. In the first case since BF and BE are 3 and 6, respectively, FE is also 3. Let $DF=x$. Since triangle FAB is similar to triangle FDE, $\frac{AF}{FD}=\frac{BF}{FE}=\frac{3}{3}$. Therefore, in this case, $AF=x$ also. Since the opposite sides of a parallelogram are equal, $BC=2x$. Since triangle BCG is similar to triangle FAG, $\frac{BG}{GF}=\frac{BC}{AF}=\frac{2}{1}$, or $BG=2\cdot GF$. Finally, since $BG+GF=BF$, which equals 3 in this case, BG is found to equal 2. Using the same method for the other cases, BG is found to equal 6 and $\frac{56}{15}$, respectively, suggesting the general relationship: $BG=\frac{(BF)(BE)}{BF+BE}$.

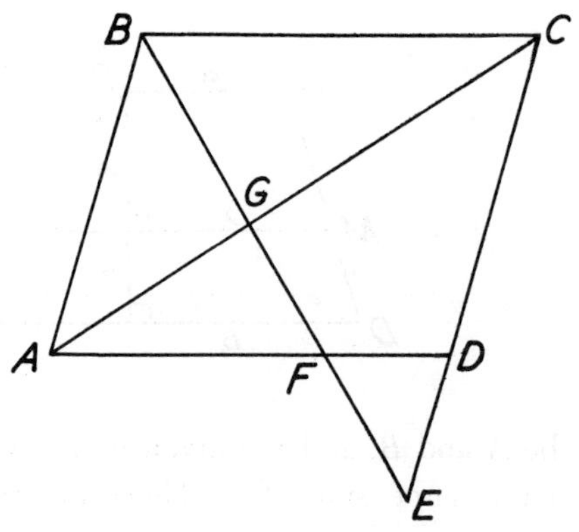

Figure 23

This relationship can readily be proved in the following manner. Since triangle BCG is similar to triangle FAG, $\frac{BG}{GF}=\frac{BC}{AF}$. Since triangle BCE is similar to triangle FAB, $\frac{BE}{BF}=\frac{BC}{AF}$. By substitution, $\frac{BG}{GF}=\frac{BE}{BF}$. Cross-multiplying and adding $(BE)(BG)$ to both members yields the equation

$$(BF)(BG)+(BE)(BG)=(BE)(GF)+(BE)(BG).$$

Factoring and rearranging terms gives:

$$BG=\frac{(BE)(GF+BG)}{BF+BE},$$

or since

$GF+BG=BF$, $BG=\frac{(BE)(BF)}{BE+BF}$. This may be written:

$\frac{1}{BG}=\frac{1}{BE}+\frac{1}{BF}$, which shows that BG is one half the harmonic mean between BE and BF. (See solution of Problem 9.)

37. Draw altitudes $\overline{CE}$ and $\overline{DF}$, as in Figures 24a and 24b. In both figures $DF=CE=\frac{1}{2}BA=\frac{1}{2}BD$. Therefore $m\angle DBA=30°$ in Fig. 24a, and $m\angle DBF=30°$ in Fig. 24b. Since in both figures, $m\angle CBA=45°$, we find $m\angle DBC=15°$ in Fig. 24a and $105°$ in Fig. 24b.

Figure 24a

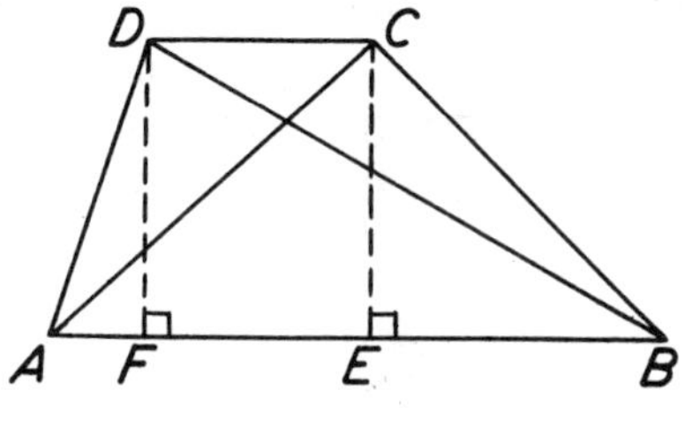

Figure 24b

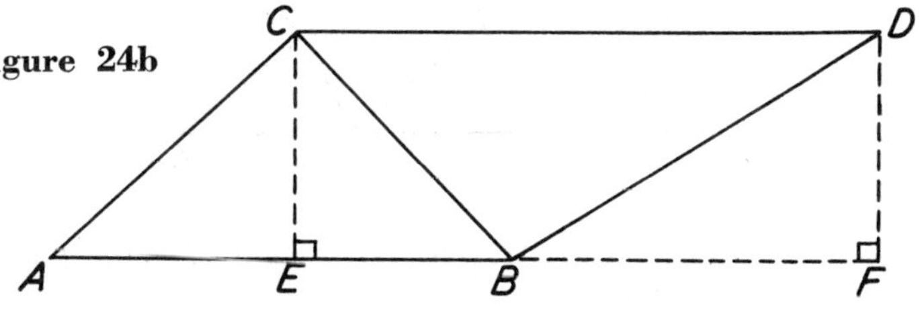

38. The diameter of each of the n^2 circles must be $\frac{12}{n}$, and each radius is $\frac{6}{n}$. Therefore the area of each circle is $\frac{36\pi}{n^2}$, and the sum of all n^2 circles is $\frac{36\pi}{n^2}(n^2)$ or 36π.

Generalizing, if the side of the square is k and the number of divisions is n, then the total area of the circles is $\frac{\pi k^2}{4}$, a result that is independent of n.

This problem can be further generalized to finding the sum of the volumes of spheres that are inscribed in a cube, or in a rectangular solid that can be divided into cubes.

Note:

A related problem is that in which the diameter of a given circle is divided into any number of parts, not necessarily equal to each other. If circles are drawn on each part as diameter, the sum of the circumferences turns out to be equal to the circumference of the given circle.

39. **(a)** Using the proportions in (**b**) it can be determined that when when $BD=(2, 3, 5)$, then $AG=(2, 3, 5)$, and $BJ=(2, 3, 5)$.

(b) Since the side parallel to one side of a triangle cuts the other two sides in proportional segments:

$$\frac{BD}{AB}=\frac{BE}{BC}=\frac{AF}{AC}=\frac{AG}{AB}=\frac{CH}{CB}=\frac{CI}{CA}=\frac{BJ}{AB}$$

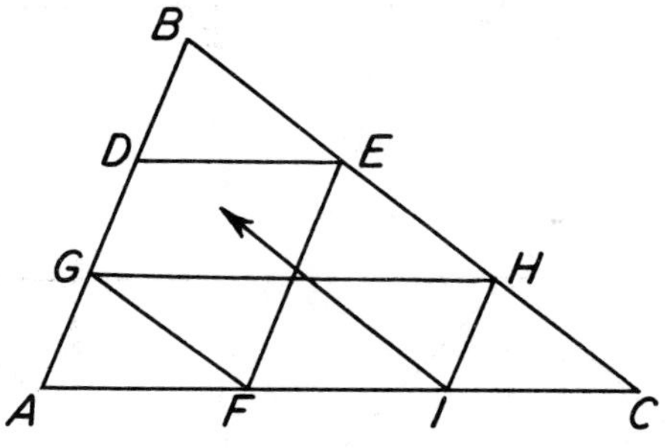

Figure 25

Hence $BD=AG=BJ$, and points D and J coincide. Also $BE=CH$ and $AF=CI$.

(c) If $BD=\frac{1}{3}BA$, then $BG=\frac{2}{3}BA$. $\overline{GH}$ divides any segment joining B to a point on $\overline{AC}$ in the ratio 2:1; this includes the median. Hence $\overline{GH}$ passes through the centroid of $\triangle ABC$ (the point of intersection of the three medians). Similarly, $\overline{EF}$ and $\overline{ID}$ pass through the centroid. Thus $\overline{GH}$, $\overline{EF}$, and $\overline{ID}$ will be concurrent when D is taken so that $BD=\frac{1}{3}BA$.

Note:

(i) If D is taken as the midpoint of $\overline{AB}$, the result will be an inscribed triangle joining the midpoints of the sides of $\triangle ABC$.

Problem

(ii) Try the same idea with any quadrilateral, pentagon, etc. (Hint: draw the line segments parallel to certain diagonals.)

40. In the diagram, if AD and DB are 2 and 3, respectively, then AF and BE are also 2 and 3, respectively. Let $FC=CE=r$; then by the Theorem of Pythagoras, $(r+2)^2+(r+3)^2=25$. Therefore, $r=1$ and the area is 6. The other area, by the same method, turns out to be 30, suggesting that the area in the general case is $(AD)(DB)$.

Figure 26

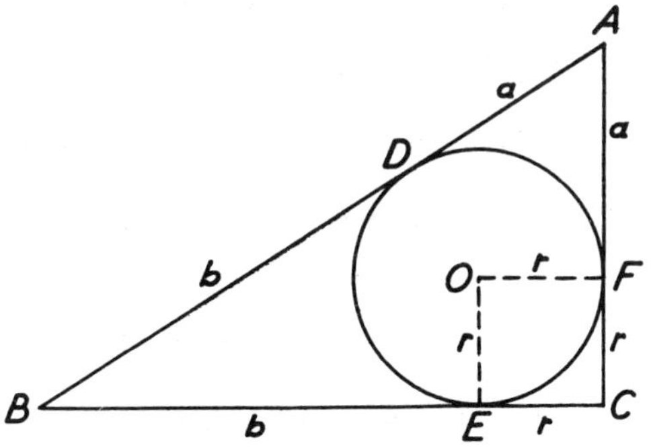

To prove this let the segments be labeled as in the diagram. Note that since $FOEC$ is a square, $\overline{FC}$ and $\overline{CE}$ may both be represented by r, the radius of the inscribed circle. The area inside the triangle is one half the product of its legs, and is also equal to rs, where s is one half the perimeter of the triangle. Therefore:

$$\frac{1}{2}(b+r)(a+r)=\frac{1}{2}r(a+b+b+r+r+a).$$

Clearing fractions and parentheses leads to: $ab=r(a+b+r)=rs$, or the area inside the triangle, as required.

41. Let c represent the length of the hypotenuse, and let 48 and a represent the lengths of the legs of the given triangle. Then $(48)^2=c^2-a^2$, or $(48)^2=(c+a)(c-a)$. If we divide $(48)^2$ into two positive factors in all possible ways, set $c+a$ equal to the larger factor, set $c-a$ equal to the smaller factor, and then solve the resulting pairs of simultaneous equations, the required solutions will be included among them.

Since $(48)^2=2^8\cdot3^2$, the 27 terms of the product $(1+2+2^2+2^3+\ldots+2^8)(1+3+3^2)$ give all its factors. Only even factors may be chosen for $c+a$ and $c-a$, since, if one were odd, the other would have to be even. The sum of $c+a$ and $c-a$, that is, $2c$, would then have to be the sum of an odd and an even number, but this is impossible. We are thus reduced to ten possibilities:

$$c+a=2^7\cdot3^2;\ 2^7\cdot3;\ 2^7;\ 2^6\cdot3^2;\ 2^6\cdot3;\ 2^6;\ 2^5\cdot3^2;\ 2^5\cdot3;\ 2^4\cdot3^2;\ 2^3\cdot3^2$$

$$c-a=2;\ 2\cdot3;\ 2\cdot3^2;\ 2^2;\ 2^2.3;\ 2^2\cdot3^2;\ 2^3;\ 2^3\cdot3;\ 2^4;\ 2^5.$$

The solutions of these taken in pairs are:

$(c, a)=(577, 575), (195, 189), (73, 55), (290, 286), (102, 90), (50, 14),$

$(148, 140), (60, 36), (80, 64)$, and $(52, 20)$.

Reference: [2], p. 111.

42. **First Solution**

If $BD>t$, then in $\triangle ABD: m\angle A>m\angle ADB$, and in $\triangle BDC$: $m\angle C>m\angle BDC$. Hence $m\angle A+m\angle C>m\angle D$, contrary to the hypothesis.

$BD<t$ can be ruled out in similar fashion. The only remaining possibility is that $BD=t$.

Figure 27

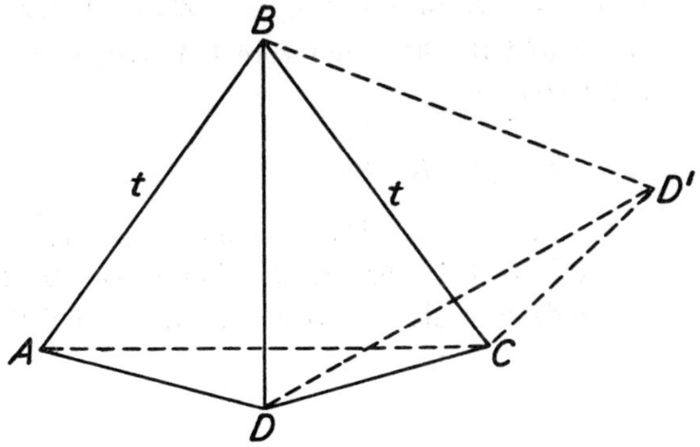

Second Solution

This is an outline of a direct proof. The missing steps are left to the reader.

Construct $\triangle BD'C$ so that $m\angle D'BC = m\angle ABD$ and $m\angle D'CB = m\angle A$, as in Fig. 27. Now prove in turn:

(1) $\triangle ABD \cong \triangle CBD'$.
(2) $\triangle ADC \cong \triangle DCD'$.
(3) $AC = DD'$.
(4) $\triangle ABC \cong \triangle DBD'$.
(5) $AB = BD = BC = BD' = t$.
(6) Specifically $BD = t$.

43. *Case 1:* $m\angle CAB < 120°$.

On $\overline{CA}$ construct equilateral triangle ECA, and draw $\overline{BE}$. Since EA, $m\angle EAB$, and AB of $\triangle EAB$ are equal, respectively, to CA,

Figure 28a

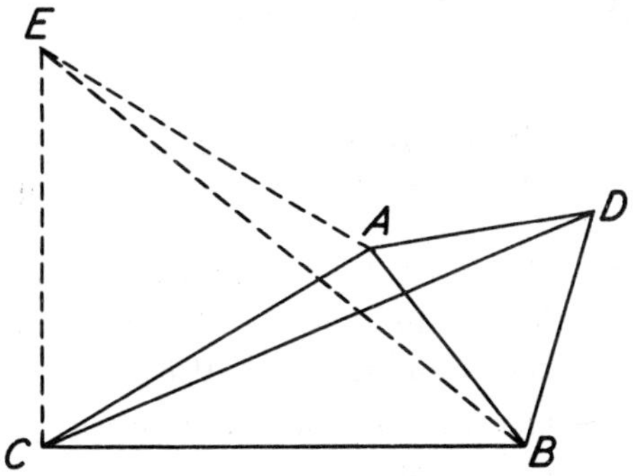

$m\angle CAD$, AD of $\triangle CAD$, $\triangle EAB \cong \triangle CAD$. Therefore $CD = EB$. But $m\angle ECB = 90°$; therefore $\overline{CA}$, $\overline{CB}$, $\overline{CD}$ can be the sides of a right triangle.

Case 2: $m\angle CAB = 120°$

In this case C, A, D are collinear (see Fig. 28b).

We have $CA = AB = DB$. Then since $\overline{DB}$, $\overline{CB}$, $\overline{CD}$ are the sides of a right triangle, so too are the sides $\overline{CA}$, $\overline{CB}$, $\overline{CD}$.

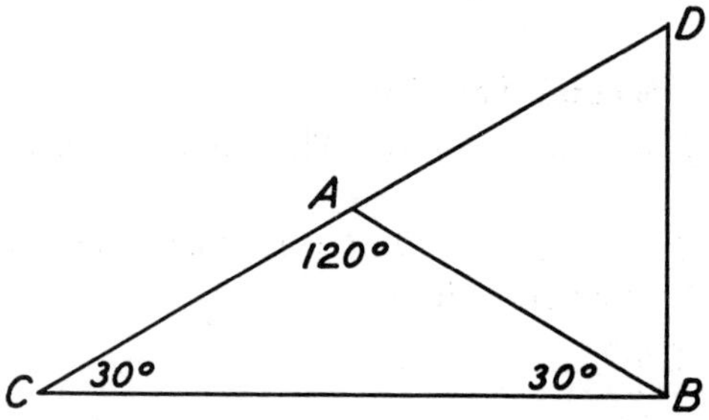

Figure 28b

Case 3: $m\angle CAB > 120°$.

The solution is left to the reader.

Related problem:

Two equilateral triangles BDC and BEC are constructed on side BC of any triangle ABC. $\overline{AD}$ and $\overline{AE}$ are drawn. Prove that $(AD)^2 + (AE)^2 = (AB)^2 + (BC)^2 + (CA)^2$. (Hint: Use the Law of Cosines.)

44. **(a)** Since $\triangle ADC$ is congruent to $\triangle ABC$, and the altitudes to $\overline{DC}$ and $\overline{CB}$ have the same length, it follows that $DC = AB$.

(b) Construct the mutual perpendicular between the two parallel lines l_1 and l_2 and call its length d. Draw a circle with Q as center and radius d. If $PQ > d$, then P lies outside of circle Q. Construct $\overleftrightarrow{PA}$ tangent to circle Q and construct $\overleftrightarrow{QW}$ parallel to $\overleftrightarrow{PA}$. These two parallel lines will form a rhombus with l_1 and l_2. A proof may be based on (**a**) above. A second rhombus usually can be obtained by constructing $\overleftrightarrow{PC}$ the other tangent to circle Q, and by constructing $\overleftrightarrow{QF}$ parallel to $\overleftrightarrow{PC}$. There will be only one solution, however, if one of the tangents $\overleftrightarrow{PA}$ or $\overleftrightarrow{PC}$ happens to be parallel to l_1 and l_2, or if $PQ=d$. There will be no solution if $PQ<d$.

45. Let $\overline{AB}$ be the hypotenuse and let $\overline{AD}$ be the sum of the legs. Draw $\overline{AD}$. At D construct a $45°$ angle, ADE. At A draw an arc with radius AB intersecting $\overline{DE}$ at B. On $\overline{AB}$ as a diameter draw a semicircle intersecting $\overline{AD}$ at C. Then ABC is the required triangle. Observe that $m\angle CDB=45°$. Hence $BC=CD$ and $AC+CB=AC+CD=AD=$ sum of the legs. Also, $\angle ACB$ is a right angle, since it is inscribed in a semicircle; $\overline{AB}$ is the hypotenuse of the desired triangle.

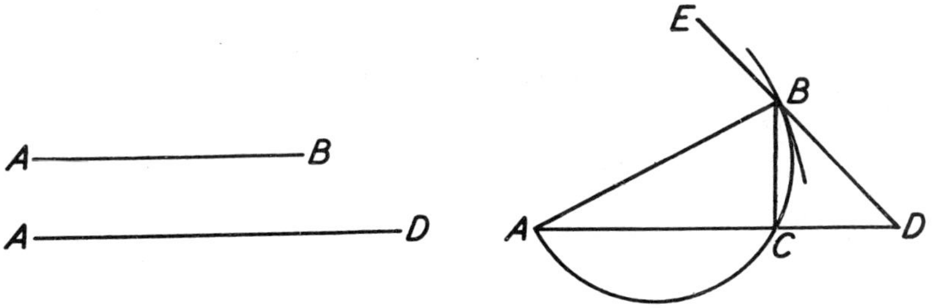

Figure 29

46. Draw $\overline{DO}$, $\overline{OE}$, $\overline{EA}$, $\overline{BE}$. Now $DE=9$.

Figure 30

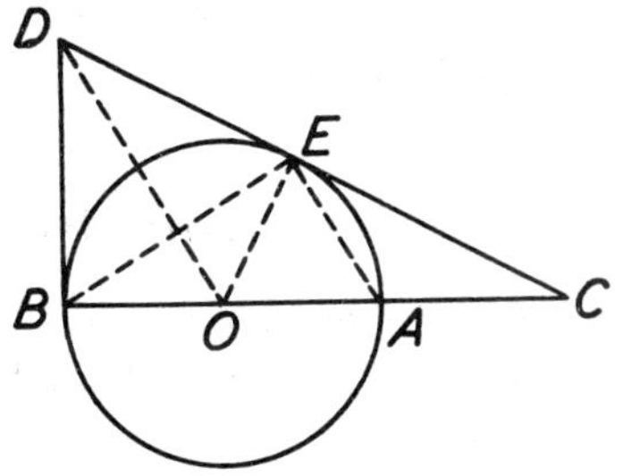

Let

$$EC=y \text{ and } BO=OA=OE=x.$$

Now $\overline{DO}$ is parallel to $\overline{EA}$ since both are perpendicular to $\overline{BE}$. Then $\frac{DE}{OA}=\frac{EC}{AC}$, or

$$\frac{9}{x}=\frac{y}{3}, \text{ or } xy=27.$$

In $\triangle OEC$, $(EC)^2=(OC)^2-(OE)^2$, or

$$y^2=(x+3)^2-x^2=6x+9.$$

Combining, $y^2+xy=6x+9+27$, or

$$y^2-36+xy-6x=0$$

which factors to give

$$(y-6)\,(x+y+6)=0.$$

Then $y=6$, and $AB=2x=9$.

47. Draw $\overline{AC}$, and let y represent its length. Since $\triangle ACD$ is a right triangle with hypotenuse $\overline{AD}$, we have $x^2=y^2+c^2$. By the Law of Cosines, $y^2=a^2+b^2-2ab\ cos\ B$. But $\cos B=-\cos D$ (opposite angles of an inscribed quadrilateral are supplementary); and $\cos D=\frac{c}{x}$. Combining these results we obtain: $x^3-x(a^2+b^2+c^2)-2abc=0$.

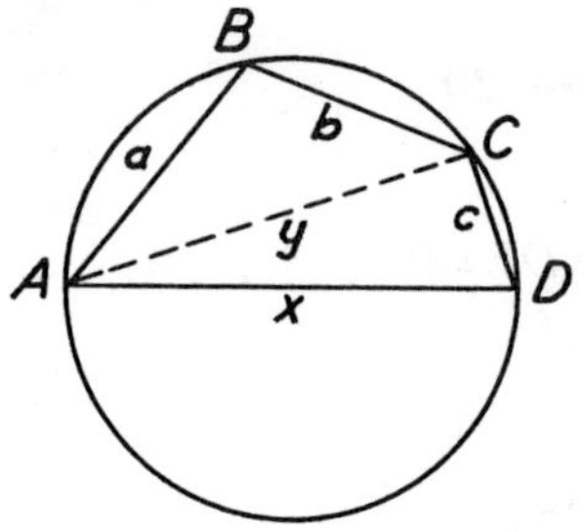

Figure 31

48. **First Method**

If $\overline{AB}$ is to be trisected (Fig. 32), draw any line $\overline{AC}$. Draw $\overline{CB}$ and extend it to D so that $CB=BD$; draw $\overline{AD}$. Bisect $\overline{AD}$ at E, and draw $\overline{CE}$. $\overline{AB}$ and $\overline{CE}$ are medians of $\triangle ACD$. Since the medians of a triangle trisect each other, $\overline{CE}$ trisects $\overline{AB}$ at G.

Figure 32

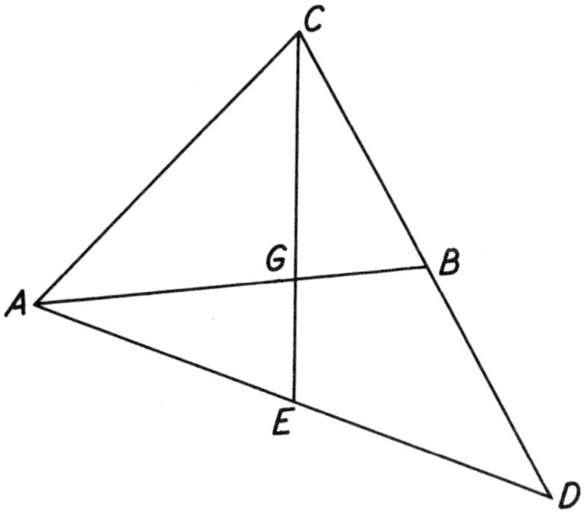

Second Method

In Fig. 33, if $\overline{CD}$ bisects $\angle C$, then $\frac{AD}{DB}=\frac{CA}{CB}$. Therefore upon $\overline{AB}$, the line to be trisected, construct any triangle ABC with $AC:CB=1:2$. Angle bisector $\overline{CD}$ will then trisect $\overline{AB}$.

Figure 33

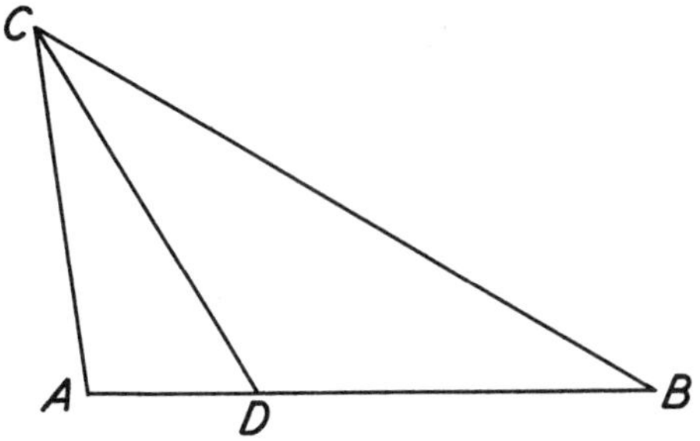

Third Method

If $\overline{AB}$ is to be trisected (Fig. 34), draw any ray $\overrightarrow{BL}$, and with any arbitrary unit of length lay off $BD=2$ units, and $DC=1$ unit. Draw $\overline{CA}$ and bisect it at E. Let $\overline{AD}$ and $\overline{BE}$ intersect at G, and let $\overrightarrow{CG}$ intersect $\overline{AB}$ at F. Then $AF=\frac{1}{3}AB$, because by Ceva's

Theorem (Appendix B), $\frac{CE}{EA}\cdot\frac{AF}{FB}\cdot\frac{BD}{DC}=1$. Hence $AF:FB=1:2$, and $AF:AB=1:3$.

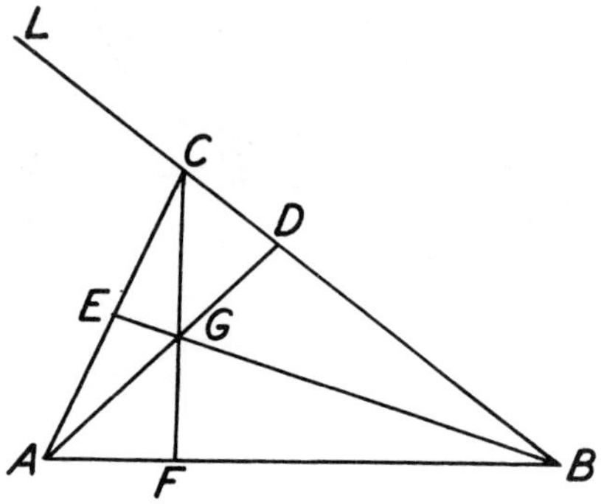

Figure 34

Fourth Method

To trisect $\overline{AB}$, construct the figure as in Fig. 35, with $CD:DA=1:5$, and with E the midpoint of $\overline{CB}$. By applying Menelaus' Theorem (Appendix C), it will be found that $AF:FB=4:1$, or $BF=\frac{1}{3}AB$.

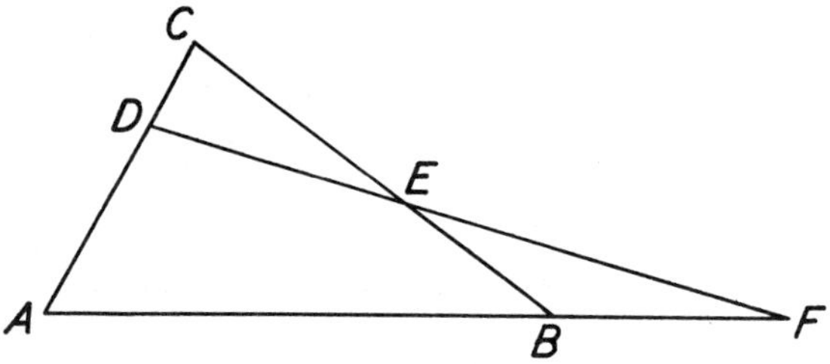

Figure 35

Problems

Devise additional methods of trisecting a line segment based on the following diagrams:

(a) $3ax=a^2$; therefore $x=\frac{a}{3}$.

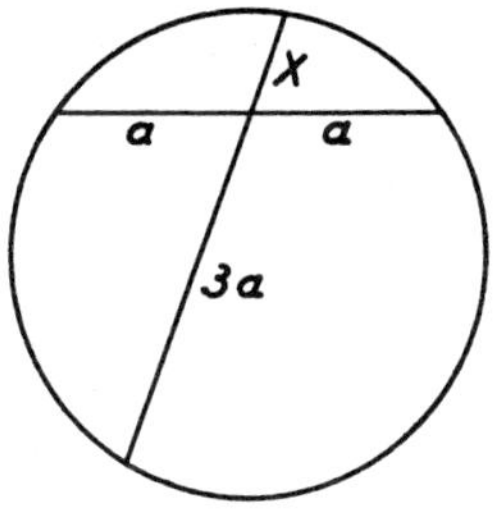

Figure 36

(b) $\frac{3a}{a} = \frac{a}{x}$

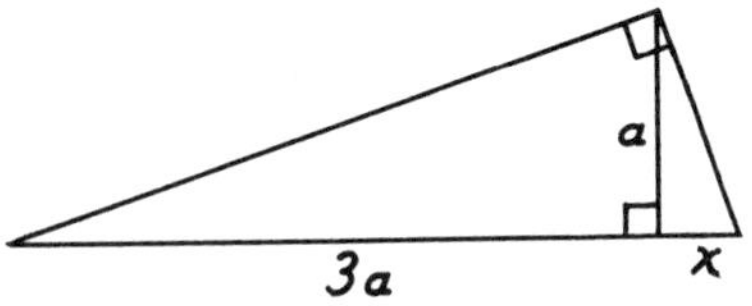

Figure 37

(c) $\frac{3a}{a} = \frac{a}{x}$

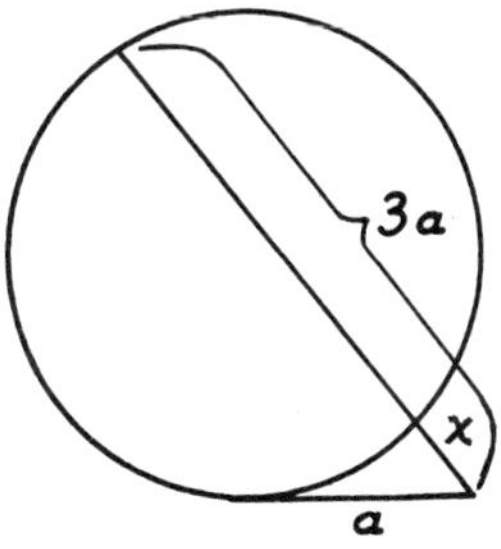

Figure 38

Note:

Of the seven methods described, all but the first lend themselves readily to the more general problem of dividing a given line segment into *any* given number of equal parts.

49. **First Solution**

Assume $m\angle 1 = m\angle 2$. Then in $\triangle AOD$, angle-bisector $\overline{OC}$ divides $\overline{AD}$ so that $\frac{AC}{CD} = \frac{OA}{OD}$. But since $AC = CD$, it follows that

$OA = OD$. This is impossible since OD cannot equal radius OA, for that would mean that D is on the circumference and chord $\overline{AB}$ would intersect the circle in three points. Hence $m\angle 1$ cannot equal $m\angle 2$.

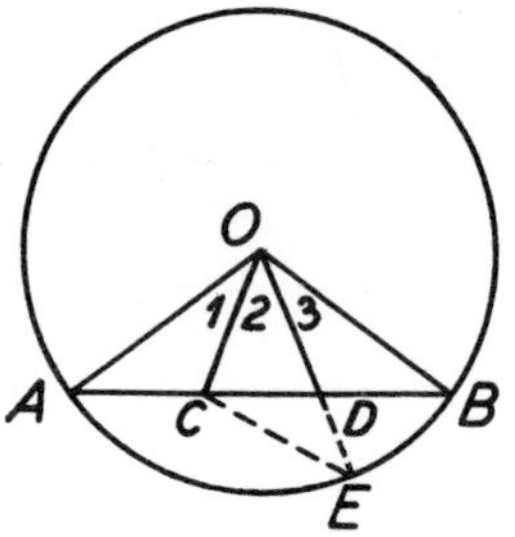

Figure 39

Second Solution

Let $\overrightarrow{OD}$ intersect the circle at E. Draw $\overline{CE}$. Then $\triangle OAC \cong \triangle OCE$ (SAS). Therefore $AC = CE$, and $m\angle E = m\angle A$. But $AC = CD$; hence $CD = CE$ and $m\angle CDE = m\angle E$ ($= m\angle A$). But if $m\angle CDE = m\angle A$, $\overline{OA} \| \overline{DE}$, contrary to the fact that they intersect at O. Therefore $m\angle 1 = m\angle 2$ is not possible.

Note:

Of the three Famous Problems of Antiquity, one that has often caught the fancy of amateur mathematicians throughout history has been that of trisecting any given angle, using only straightedge and compasses. Perhaps this is so because the problem looks so easy to solve. The fact that it has been established mathematically (about 1800) that the construction is impossible, seems to have invited, rather than deterred, amateurs to seek a solution. High school students have not been immune to this infection. Invariably when the writer has pointed out in his classroom, the impossibility of a general solution (a specific angle such as 90° or 180° can, of course, be trisected within the stated rules), one or more students would bring in an alleged solution based on the diagram of this problem.

Methods have been devised which appear to accomplish this construction; but in each case there is a violation (sometimes subtly disguised) of the limitation to straightedge and compasses. One such construction is as follows.

Problem: to trisect $\angle QOP$, (Fig. 40).

Figure 40

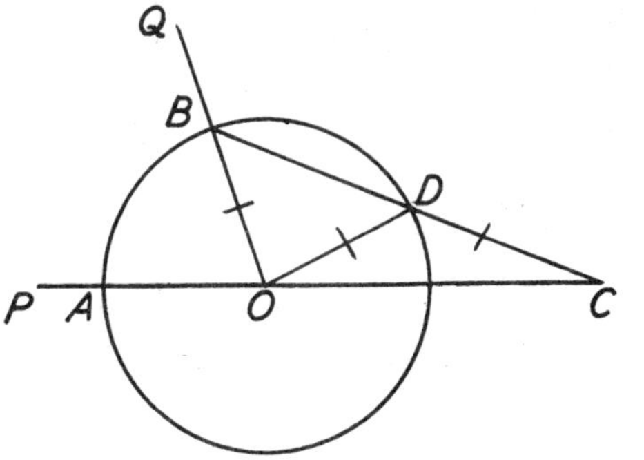

Draw a circle with O as center, intersecting $\overrightarrow{OP}$ at A and $\overrightarrow{OQ}$ at B. Open the compasses a distance equal to the radius of the circle, place the two points on a straightedge, and manipulate the latter so that it passes through B, and so that the two points of the compasses fall on the circumference D and on $\overrightarrow{PO}$ at C. This makes $\overline{DC}$ equal in length to the radius of the circle$=OD=OB$. If $m\angle DCO=x$, then $m\angle DOC=x$. Also, since $\angle BDO$ is an exterior angle of $\triangle ODC$, $m\angle BDO=2x$. Therefore $m\angle OBD=2x$, and $m\angle POQ=3x=(m\angle C+m\angle CBO)$. Thus $m\angle DCO=\frac{1}{3}m\angle QOP$, which has thus been trisected.

The above proof is correct, but the construction does not meet the requirement that a straightedge be used. A straightedge is an instrument for drawing a straight line and must be unmarked. The two compass points have the effect of marking it.

References: [26], [52], [53], Chapter 8.

50. **First Proof**

If K is the midpoint of $\overline{AB}$, draw $\overline{DK}$ intersecting $\overline{AE}$ at M. Then $BK=FD$. Hence $BFDK$ is a parallelogram and $\overline{KD}$ is parallel to $\overline{BF}$. In $\triangle AEB$, $\overline{KM}$ is parallel to $\overline{BE}$ and bisects $\overline{AB}$. Therefore $\overline{KM}$ bisects $\overline{AE}$. Thus we have shown that $\overline{DM}$ is the perpendicular bisector of $\overline{AE}$. Therefore $DE=AD$.

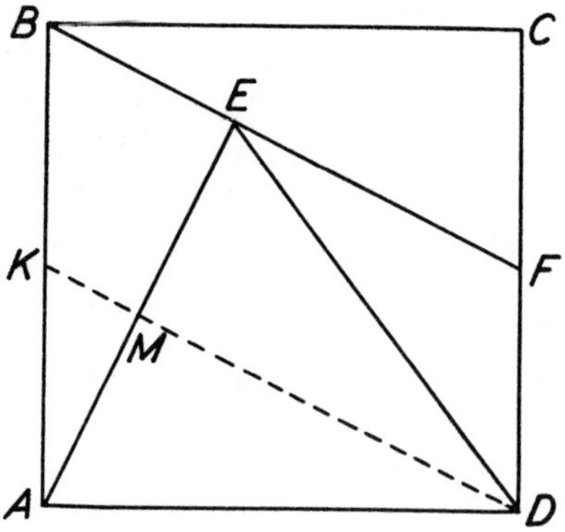

Figure 41

Second Proof

Construct square $CHGD$ next to $ABCD$. Diagonal $\overline{BG}$ of rectangle $BHGA$ passes through the midpoint of $\overline{CD}$ which is F. $\triangle AEG$ is a right triangle, and $\overline{ED}$ is the median to hypotenuse $\overline{AG}$. Therefore $ED = \frac{1}{2}AG = AD$.

Third Proof

Draw $\overline{AF}$. Since $\overline{AE}$ is perpendicular to $\overline{BF}$ and $\overline{FD}$ is perpendicular to $\overline{AD}$, A, E, F, D are concyclic (Appendix E). Therefore (1) $m\angle AED = m\angle AFD$ (both subtend the same arc AD). Now $m\angle EAD = m\angle EFC$ (both are supplementary to $\angle EFD$). Also $m\angle EFC = m\angle AFD$ ($\triangle BCF \cong \triangle AFD$). Hence (2) $m\angle EAD = m\angle AFD$. From (1) and (2) it follows that $m\angle AED = m\angle EAD$. Therefore $AD = DE$.

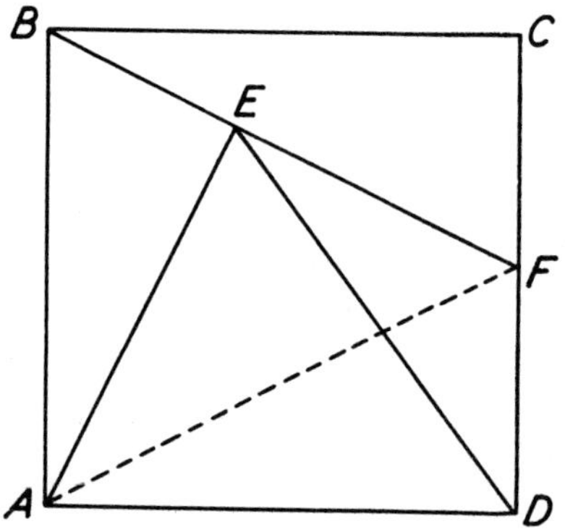

Figure 42

Fourth Proof

Let $\overleftrightarrow{AD}$ and $\overleftrightarrow{AB}$ be the x and y axes, respectively, of a rectangular coordinate system. If the coordinates of D are taken as $(2a, 0)$, then those of B, C, and F are $(0, 2a)$, $(2a, 2a)$, and $(2a, a)$ respectively. The coordinates of E are determined by solving two equations, viz. that of $\overleftrightarrow{BF}$, and that of the line through A perpendicular to $\overline{BF}$. Finally it can be determined by the distance formula that $DE = DA$.

Note:

The first proof will hold for parallelogram $ABCD$, shown in Fig. 43. Therefore the problem is valid for any parallelogram, and therefore for any rectangle.

Figure 43

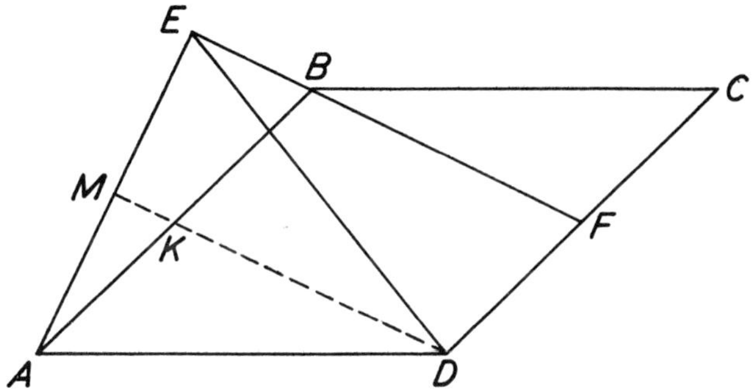

51.

First Proof

Construct an equilateral triangle outward on side $\overline{BC}$. This makes $\overline{BE} \| \overline{CD}$. Therefore $\frac{BC}{CD} = \frac{BE}{CD} = \frac{AE}{AC} = \frac{AC+CE}{AC} = \frac{AC+CB}{AC} = 1 + \frac{BC}{AC}$. Divide through by BC: $\frac{1}{CD} = \frac{1}{BC} + \frac{1}{AC}$.

Figure 44

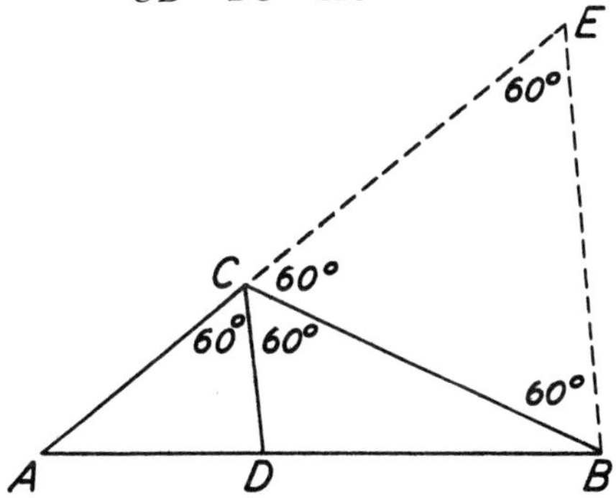

Second Proof

Area $\triangle ACB$ = area $\triangle ACD$ + area $\triangle DCB$. Therefore $\frac{1}{2}$ (CA) (CB) $\sin 120° = \frac{1}{2}(CA)(CD) \sin 60° + \frac{1}{2}(CD)(CB) \sin 60°$. But $\sin 120° = \sin 60°$; therefore, if we divide through by $\frac{1}{2}(CA)(CB)(CD) \sin 60°$, we obtain the required result.

Third Proof

If equilateral triangle ACF is constructed outwardly on side AC in Fig. 45, then $\overline{AF} \| \overline{CD} \| \overline{BE}$, and the required result follows at once from Problem 30.

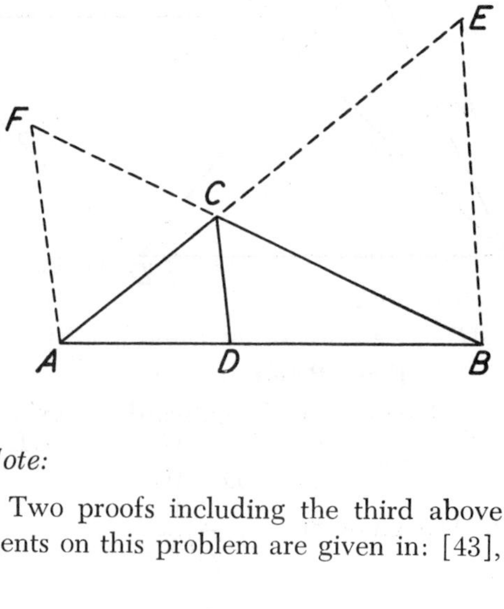

Figure 45

Note:

Two proofs including the third above and additional comments on this problem are given in: [43], vol. 12, 1910/3, p. 37.

52.

First Solution

For any position of J, $\triangle BJE \cong \triangle EKC$ ($BE = EC$; $\angle BEJ$ and $\angle KEC$ are each complements of $\angle JEC$; $\angle EBJ \cong \angle ECK$); hence $BJ = KC = a$, where a is chosen arbitrarily. Quadrilateral $EJCK$ can always be divided into triangles JEC and ECK which have base lengths of $12 - a$ and a, respectively. Since the altitude of both triangles is 6, area $\triangle JEC$ + area $\triangle ECK = \frac{6}{2}(12-a) + \frac{6}{2}a = 3(12 - a + a) = 36$, which is one quarter of the area of the given square.

Figure 46

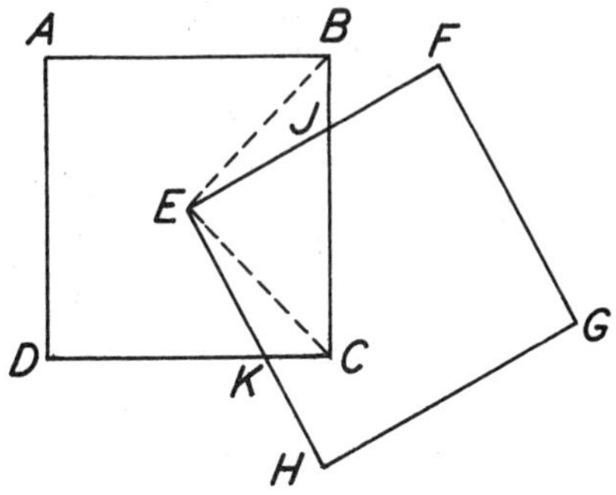

Second Solution

If $\overrightarrow{JE}$ and $\overrightarrow{KE}$ are prolonged to meet $\overline{AD}$ and $\overline{AB}$ at N and M respectively, four congruent quadrilaterals are formed. Hence $JECK$, which is one of them, is one quarter of the area of the square.

Note:

A further generalization can be made for any regular polygon. This is suggested by Figs. 47 and 48. In each figure, it is the central angle AEC that rotates to form quadrilateral $EJCK$, whose area is constant and which has the same measure as the area of $\triangle AEC$. The proof is similar to that of the first solution above.

Figure 47

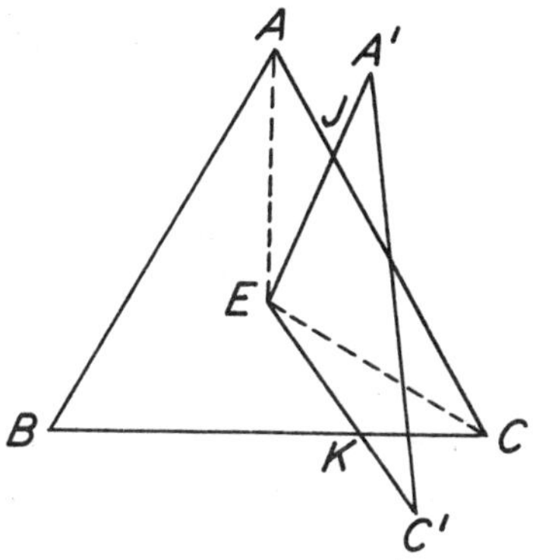

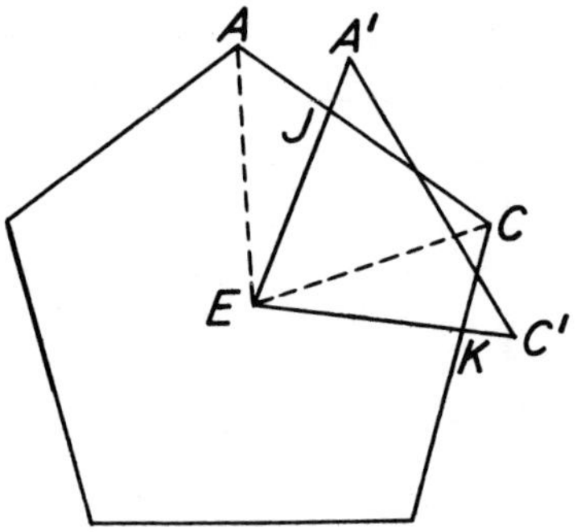

Figure 48

53. If chord $\overline{KJ}$ is one side of the proposed heptagon, then by the Law of Cosines: $\cos x = \dfrac{r^2 + r^2 - \dfrac{3r^2}{4}}{2r^2} = \dfrac{5}{8} = .625$. Hence $x \approx 51°19'5''$. But the central angle of a heptagon should be $\dfrac{360°}{7}$ or about $51°25'43''$, a difference which is not sufficient to be noticed in an ordinary compass construction.

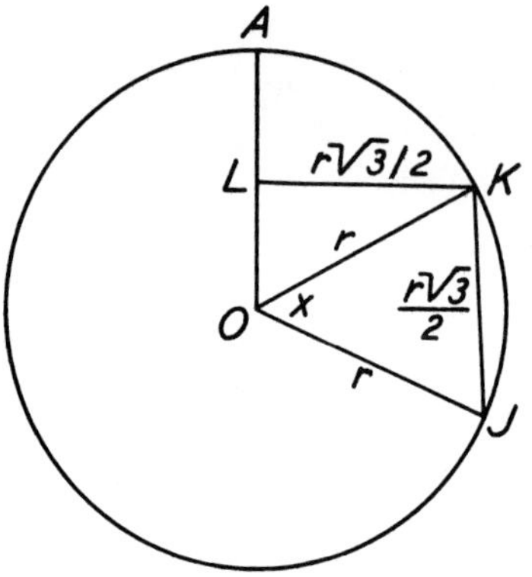

Figure 49

Notes:

(i) Solver Allen Flusberg (Brooklyn, New York) discovered that a good approximation of a regular nonagon (nine-sided polygon) can be constructed by using the fact that $\sin 40°$ (≈ 0.64279) is very close to $\dfrac{9}{14}$ (≈ 0.64286). If a right triangle be constructed with hypotenuse 14 and leg 9, the measure of the angle opposite 9 may be used as the central angle of the nonagon.

(ii) At the age of 17, Gauss (1777-1855) proved that a regular polygon of N sides is constructible with only straightedge and compasses if and only if N is of the form $2^{n} \cdot p_1 \cdot p_2 \ldots$, Where $p_1, p_2, \ldots$ are distinct primes of the form $2^{2^n}+1$.

For $n=0, 1, 2, 3, 4$, $2^{2^n}+1=3, 5, 17, 257, 65{,}537$, all of which are primes. Fermat (1601-1665), possibly the greatest of amateur mathematicians, conjectured that $2^{2^n}+1$ was a prime for every integral value of n. He was shown to be mistaken by Euler (1707-1783), who showed in 1732 that $2^{2^5}+1=641 \cdot 6{,}700{,}417$. To date, only the above five primes have been found of the form $2^{2^n}+1$.

(iii) In the following short list, the underlined numbers correspond to inscriptible regular polygons. The reader may verify and extend the list by using Gauss' formula given in note (**ii**).

$$N = \underline{3}, \underline{4}, \underline{5}, \underline{6}, 7, \underline{8}, 9, \underline{10}, 11, \underline{12}, 13, 14, \underline{15}, \underline{16}, \underline{17}, 18, 19, \underline{20}, 21, 22, 23, \underline{24}.$$

References: [26] and [53], Chapter 8.

54. Because the lengths of tangents from an exterior point to a circle are equal, we may designate the segments as in Fig. 50.

Figure 50

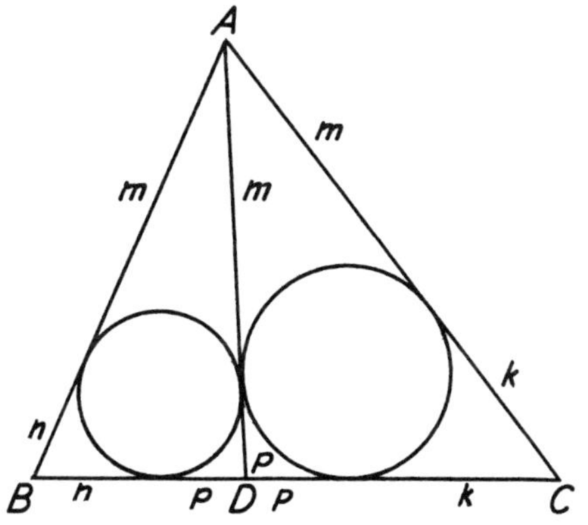

We have

$$2m+2k+2n+2p=a+b+c;$$

$$m+k=b$$

and

$$n+p=x.$$

Hence

$$2b+2x=a+b+c,$$

and

$$BD=x=\frac{a-b+c}{2}$$

For the circle inscribed in the entire triangle, designate the lengths of tangents as in Fig. 51.

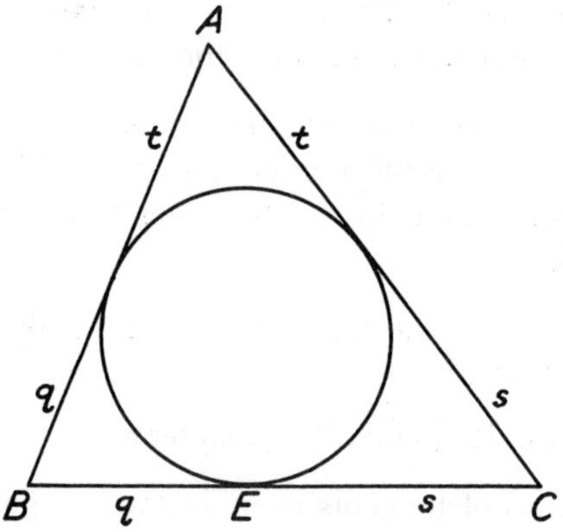

Figure 51

Here

$$2q+2s+2t=a+b+c$$

and

$$s+t=b.$$

Hence

$$2q+2b=a+b+c.$$

Therefore

$$BE=q=\frac{a-b+c}{2}=x.$$

Thus E and D coincide, making $DE=0$.

Note:

This problem may be presented as one of construction: Determine point D on side $\overline{BC}$ of $\triangle ABC$, so that the circles inscribed in triangles ABD and ADC are tangent to $\overline{AD}$ at the same point. D is determined as the point of tangency of the inscribed circle of $\triangle ABC$ with side $\overline{BC}$.

55. Extend $\overrightarrow{EA}$ to meet $\overline{DF}$ at H. This is possible because $\overline{EA} \| \overline{BC}$, which intersects $\overline{DF}$. Therefore $m\angle 2 = m\angle 3 = m\angle 8$; and $180° = m\angle 4 + m\angle 5 + m\angle 8 = m\angle 4 + m\angle 5 + m\angle 2$. Now $m\angle 7 + m\angle 1 + m\angle 6 = 180°$ (opposite angles of an inscribed quadrilateral). Therefore $m\angle 4 + m\angle 5 + m\angle 2 = m\angle 7 + m\angle 1 + m\angle 6$. But $m\angle 7 = m\angle 4$, ($\overline{AB} \| \overline{CF}$), and $m\angle 5 = m\angle 6$, ($CF = CD$). Hence $m\angle 1 = m\angle 2$; from which it follows that $AD = AH$. But since $AD = AE$, median AD of $\triangle DEH$ equals one-half of EH. Therefore $\triangle EDH$ is a right triangle, with $\overline{DE}$ perpendicular to $\overline{DH}$.

Figure 52

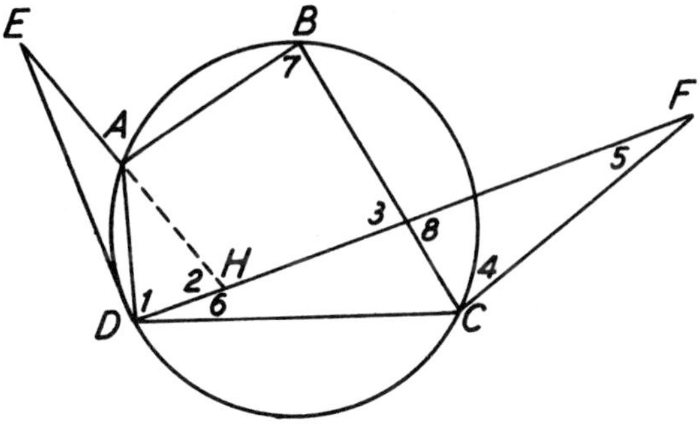

Note:

For readers familiar with the elements of vector analysis, the following proof may be of interest:

Let vector DA (represented by $\overrightarrow{DA}$) equal $a\boldsymbol{\alpha}$, where $\boldsymbol{\alpha}$ is the unit vector in the director of $\overrightarrow{DA}$, and a is the scalar multiplier. Similarly, let $\overrightarrow{AB} = b\boldsymbol{\beta}$, $\overrightarrow{CB} = c\boldsymbol{\gamma}$, $\overrightarrow{DC} = d\boldsymbol{\delta}$. Then $\overrightarrow{DE} = a\boldsymbol{\alpha} + a\boldsymbol{\gamma}$; $\overrightarrow{DF} = d\boldsymbol{\delta} + d\boldsymbol{\beta}$. Therefore the scalar product $(\overrightarrow{DE}) \cdot (\overrightarrow{DF}) = (a\boldsymbol{\alpha} + a\boldsymbol{\gamma}) \cdot (d\boldsymbol{\delta} + d\boldsymbol{\beta}) = ad(\boldsymbol{\alpha} \cdot \boldsymbol{\delta} + \boldsymbol{\alpha} \cdot \boldsymbol{\beta} + \boldsymbol{\gamma} \cdot \boldsymbol{\delta} + \boldsymbol{\gamma} \cdot \boldsymbol{\beta})$. But $\boldsymbol{\alpha} \cdot \boldsymbol{\delta} + \boldsymbol{\beta} \cdot \boldsymbol{\gamma} =$ the sum of the cosines of the opposite angles of an inscribed quadrilateral, which equals the sum of the cosines of a pair of supplementary angles, which equals zero. Similarly, $\boldsymbol{\alpha} \cdot \boldsymbol{\beta} + \boldsymbol{\gamma} \cdot \boldsymbol{\delta} = 0$. Therefore $\overline{DE}$ is perpendicular to $\overline{DF}$.

56. If the triangle is equilateral, its area is $\frac{s^2\sqrt{3}}{4}$, an irrational number, because $s^2 = (x_2 - x_1)^2 + (y_2 - y_1)^2$ is a rational number. If it can be shown that the area must also be rational, the contradiction would lead to the conclusion that the triangle cannot be equilateral.

First Method

The area of the triangle ABC is rational because it can be expressed as the difference between the area of the circumscribed rectangle and the sum of the areas of the right triangles. Since the vertices of the rectangle and of the right triangles are integers, their areas are rational. Hence the area of $\triangle ABC$ is rational.

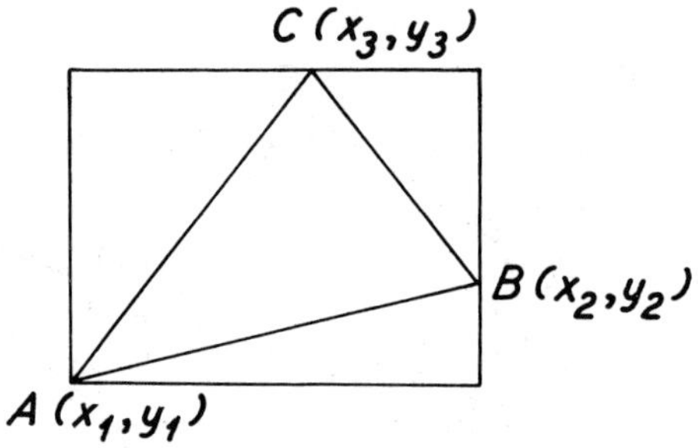

Figure 53

Second Method

The area of $\triangle ABC$ may be exhibited as a determinant:

$$\text{Area} = \frac{1}{2} \begin{vmatrix} x_1 & y_1 & 1 \\ x_2 & y_2 & 1 \\ x_3 & y_3 & 1 \end{vmatrix}$$

which is rational because the coordinates are integers.

57. Since triangles BJG and CJA are similar and since $BG = CB$, $\frac{BJ}{CJ} = \frac{CB}{CA}$. For similar reasons, $\frac{CH}{AH} = \frac{CB}{CA}$. Therefore, $\frac{BJ}{CJ} = \frac{CH}{AH}$. Since $CJ = CB - BJ$, and $AH = CA - CH$, we have $\frac{BJ}{CB - BJ} = \frac{CH}{CA - CH}$. Hence $BJ \cdot CA - BJ \cdot CH = CB \cdot CH - BJ \cdot CH$, and $\frac{BJ \cdot CA}{2} = \frac{CB \cdot CH}{2}$.

Figure 54

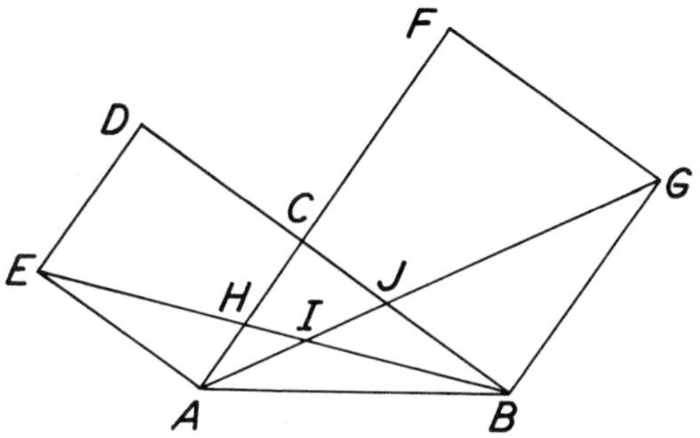

Thus, area of triangle BJA = area of triangle CHB. Now if from each of these we subtract triangle JIB, we obtain the required result.

This result can be generalized for any triangle, if rhombuses $AEDC$ and $CFGB$ are constructed as shown in Fig. 55*a*.

Draw a line through J parallel to $\overline{CA}$, intersecting $\overline{AB}$ at K. Draw a line through H parallel to $\overline{BC}$, intersecting $\overline{AB}$ at K'. If it can be proven that $AK'+KB=AB$, then K and K' must be the same point, and $CJKH$ is a parallelogram.

Figure 55a

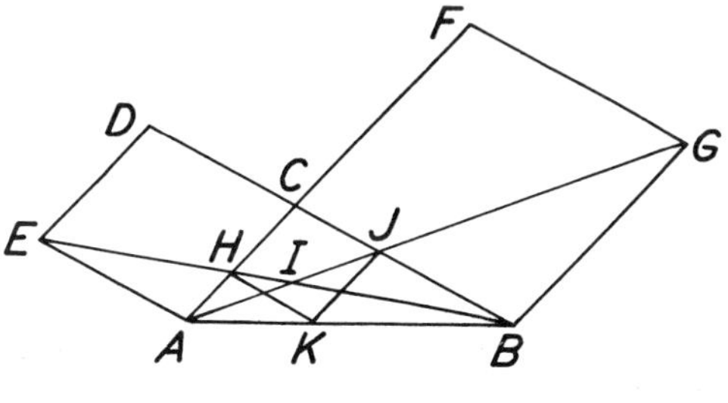

Let BC, CA, AB be represented by a, b, c, respectively. By similar triangles, we have: $\frac{BK}{BA}=\frac{BJ}{BC}$; also $\frac{BJ}{JC}=\frac{BG}{CA}=\frac{a}{b}$. Hence $\frac{BJ}{BJ+JC}=\frac{a}{a+b}$ (Appendix A). Therefore $BK=\frac{ca}{a+b}$. Similarly, it can be shown that $AK'=\frac{cb}{a+b}$. Hence $BK+K'A=c=AB$, so that K' coincides with K, and $CJKH$ is a parallelogram.

Now area $\triangle JAK$ = area $\triangle JCK$ (both have base $\overline{JK}$, and altitudes are equal); and area $\triangle HKB$ = area $\triangle HKC$ (similar reason). Therefore area $\triangle JAK$ + area $\triangle HKB$ = area $CHKJ$. Now if area $HIJK$ is subtracted from each, the required result is obtained.

Although it is not needed in the above proof, $CJKH$ can be shown to be a rhombus. If Fig. 55*b* is extracted from Fig. 55*a*, it follows from Problem 30 that: $\frac{1}{HK}=\frac{1}{EA}+\frac{1}{BC}=\frac{1}{a}+\frac{1}{b}$.

Similarly, it can be shown that: $\frac{1}{KJ}=\frac{1}{CA}+\frac{1}{BG}=\frac{1}{a}+\frac{1}{b}$.

Therefore $HK=KJ$, and $CJKH$ is a rhombus.

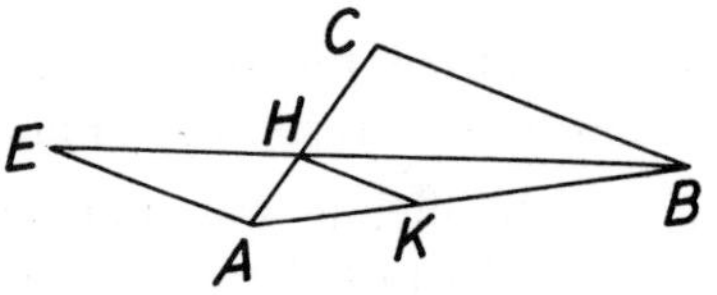

Figure 55b

Note:

The figure consisting of two squares constructed on two sides of a triangle (not necessarily a right triangle) lends itself to many problems. In addition to the results in Problem 57, here is a selection of such problems. Hints for solutions are given at the end of the list.

(1) In Fig. 54, **prove:** $(CH)^2=(CJ)^2=(AH)(JB)$.

(2) In Fig. 54, **prove** that the altitude from C, $\overline{AJ}$ and $\overline{BH}$ are concurrent.

(3) In Fig. 54, if altitude $\overline{CP}$ intersects $\overline{AG}$ in Q, **prove:**

(a) $\frac{1}{CQ}=\frac{1}{AB}+\frac{1}{CP}$.

(b) CQ equals the length of the side of the square inscribed in $\triangle ABC$ with one side lying along $\overline{AB}$.

(4) **Given:** $EBAD$ and $BCGF$ are squares on sides $\overline{AB}$ and $\overline{BC}$ of $\triangle ABC$; $\overline{BH}\perp\overline{AC}$.

Prove: $EJ=JF$.

Figure 56

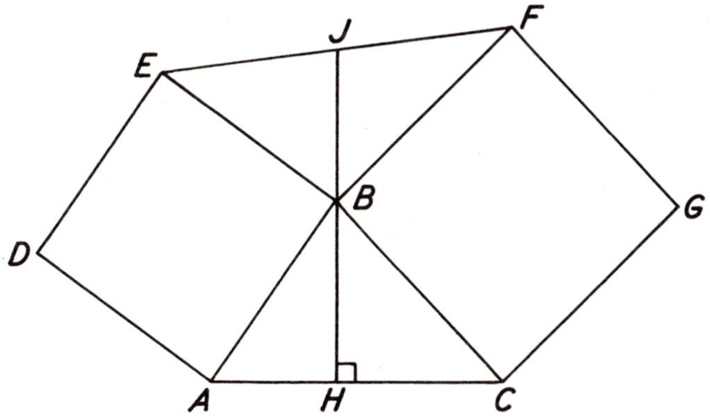

(5) Prove the converse of (4).

(6) **Given:** $ABFG$ and $BCDE$ are squares on sides $\overline{AB}$ and $\overline{BC}$ of $\triangle ABC$; $\overline{AF}$, $\overline{GB}$, $\overline{BD}$, and $\overline{EC}$ are diagonals; $AM = MC$.

Prove: (a) $MH = MI$

(b) $\overline{MH} \perp \overline{MI}$

Figure 57

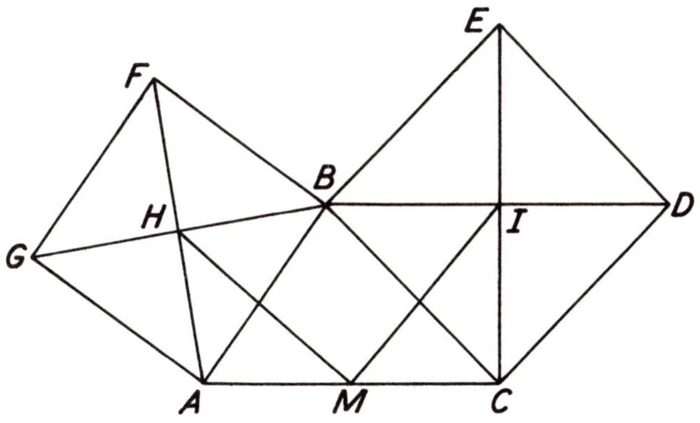

(7) **Given:** Rt. $\triangle ABC$, with rt. angle at A; squares $ABDE$, $ACGF$, and $BCKL$ on sides $\overline{AB}$, $\overline{AC}$, and $\overline{BC}$.

Prove: $(DL)^2 + (GK)^2 = 5(BC)^2$.

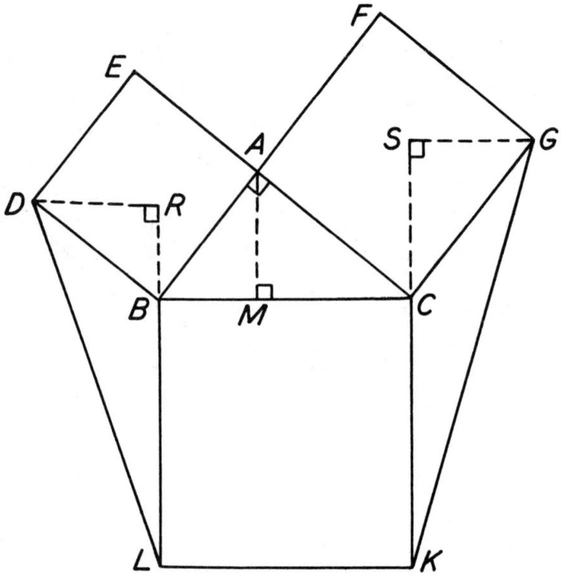

Figure 58

(8) Problem 64 is essentially based on two squares on two sides of a triangle. The triangle is GFE, and the squares are incomplete: GPA__ and GEB__.

HINTS FOR SOLUTIONS

(1) This follows readily from the proof of Problem 57.

(2) Apply Ceva's Theorem (Appendix B).

(3) **(a)** Apply Menelaus' Theorem (Appendix C). **(b)** See the second method in the solution of Problem 33.

(4) Draw $\overline{EQ}\perp\overleftrightarrow{JB}$ and $\overline{FR}\perp\overleftrightarrow{JB}$, with Q and R on $\overleftrightarrow{JB}$. Show that $\overline{EQ}$ and $\overline{FR}$ are each congruent to $\overline{BH}$ by proving appropriate triangles congruent.

(5) Apply the indirect method, using the result in Problem 4, above.

(6) Draw $\overline{FC}$ and $\overline{AE}$ intersecting at Q.

(a) Prove $\triangle FBC \cong \triangle ABE$, so that $FC = AE$. Then $HM = \frac{1}{2}FC$, etc.

(b) Show that F, B, Q, A are concyclic (Appendix E). Therefore $m\angle FQA = m\angle FBA$, etc.

(7) Prolong $\overrightarrow{LB}$ and draw $\overline{DR} \perp \overrightarrow{LB}$, with R on $\overrightarrow{LB}$. Prolong $\overrightarrow{KC}$ and draw $\overline{GS} \perp \overrightarrow{KC}$, with S on $\overrightarrow{KC}$. Draw altitude $\overline{AM}$ with M on $\overline{BC}$. Prove $\triangle DRB \cong \triangle ABM$ and $\triangle GCS \cong \triangle AMC$. Determine corresponding sides of these triangles. Apply the Theorem of Pythagoras to $\triangle LRD$ and $\triangle KSG$: $(LD)^2 = (DR)^2 + (RB + BL)^2$; $(KG)^2 = \ldots$, and make appropriate substitutions. Also make use of: $(BM)(MC) = (AM)^2$.

58. $(AB)^2 = (AE)^2 = (AC)(AD)$. Therefore $\frac{AC}{AE} = \frac{AE}{AD}$, whence $\triangle ACE \sim \triangle ADE$. Hence $m\angle CEA = m\angle D$. But $m\angle D = m\angle GFE$ (each is supplementary to angle GFC). Therefore $m\angle CEA = m\angle GFE$, and $\overline{GF} \| \overline{AE}$.

If E is inside the circle, the proof is the same except for the statement in the parentheses.

59. **First Solution**

Construct equilateral $\triangle AGD$ on $\overline{AD}$ as base, and draw $\overline{EG}$. Then since $EA = ED$ and $GA = GD$, $\overline{EG}$ is the perpendicular bisector of $\overline{AD}$. Hence, $\triangle EDC \cong \triangle EGD$, $(GD = DC$, $m\angle GDE$ $(=75°) = m\angle EDC$, $DE = DE)$.

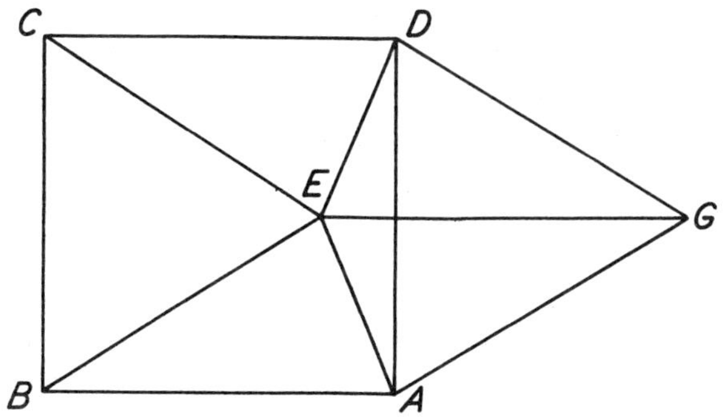

Figure 59

Therefore $m\angle CED = m\angle DEG = 75° = m\angle CDE$; and $CD = CE$. But $CD = CB$; therefore $CE = CB$. Similarly, it can be proved that $BE = CB$, so that $\triangle BCE$ is equilateral.

Second Solution

On $\overline{BA}$ construct $\triangle ABF$ with base angles both equal in measure to $15°$, as in Fig. 60.

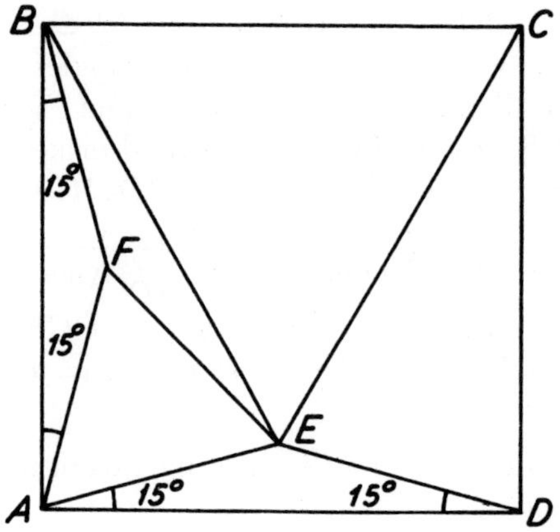

Figure 60

Then $\triangle ABF \cong \triangle AED$, and $AF = AE$. Since $m\angle FAE = 60°$, $\triangle AFE$ is equilateral. Therefore, $FE = FA = FB$.

Since $m\angle BFA = 150°$, and $m\angle AFE = 60°$, it follows that $m\angle BFE = 150°$, and $m\angle FBE = 15°$. Thus, $m\angle EBC = 60°$. Similarly, it can be proved that $m\angle BCE = 60°$, making $\triangle BEC$ equilateral.

Third Solution

Construct an equilateral triangle on $\overline{BC}$ as base, with vertex E' inside the square. The plan is to show that E' coincides with E.

Since $BE' = BC = BA$, $\triangle ABE'$ is isosceles, $m\angle ABE' = 30°$ and $m\angle BAE' = 75°$. Therefore E' lies on $\overrightarrow{AE}$. Similarly it can be be proven that E' lies on $\overrightarrow{DE}$. Hence E' and E coincide.

Note:

Solutions may also be based on the use of trigonometric formulas. For example, it may be shown that the altitude from E of isosceles triangles BEC has a length which can only be that of an equilateral triangle.

60.

First Solution

Draw $\overline{RS}$ perpendicular to $\overline{AD}$ and $\overline{GR}$ and $\overline{FS}$ perpendicular to $\overline{RS}$, as in Fig. 61. Let $2s$ represent the length of each side of the hexagon, and x the length of $\overline{AP}$. Therefore $AG=s$, $AF=FE=2s$, and $AD=4s$ (the length of a side of a regular hexagon is equal to the length of its radius).

Figure 61

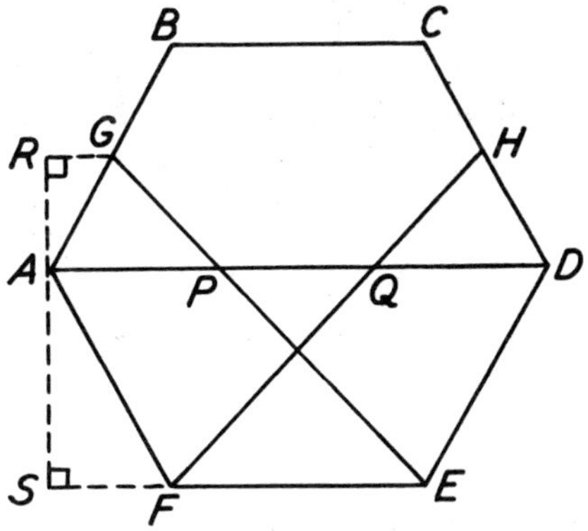

Since $m\angle GAP=m\angle PAF=60°$, then $m\angle RAG=m\angle SAF=30°$. Therefore $RG=\frac{s}{2}$, $RA=\frac{s\sqrt{3}}{2}$, $SF=s$, $AS=s\sqrt{3}$. Now $RGES$ is a trapezoid with $\overline{AP}$ parallel to its bases. Therefore by the second formula in the solution of Problem 20,

$$x=\frac{AS\cdot RG+AR\cdot SE}{AS+AR}=\frac{s\sqrt{3}\left(\frac{s}{2}\right)+3s\left(\frac{s\sqrt{3}}{2}\right)}{s\sqrt{3}+\frac{s\sqrt{3}}{2}}=\frac{4s}{3}.$$

Similarly, $QD=\frac{4s}{3}$. And finally, $PQ=AD-(AP+QD)=\frac{4s}{3}$.

Second Solution

Let O be the center of the polygon, and I the midpoint of $\overline{ED}$. Draw $\overline{AI}$ and $\overline{GI}$. $\overline{AB}\|\overline{DE}$; therefore the line through O perpendicular to both of them bisects both chords and must coincide with $\overline{GI}$. Since $AB=DE$, $AG=IE$, and $\overline{AO}$ is a median of $\triangle AGI$. Now $AGIE$ is a parallelogram since $AG=IE$; $\overline{AG}\|\overline{IE}$ ($AGIE$ is actually a rectangle). Hence diagonal $\overline{GE}$ bisects $\overline{AI}$ at M, and $\overline{GM}$ is a second median of $\triangle AGI$. But medians $\overline{AO}$ and $\overline{GM}$ trisect each other at P. Therefore $AP=\frac{2}{3}AO=\frac{1}{3}AD$. The rest follows as in the first solution.

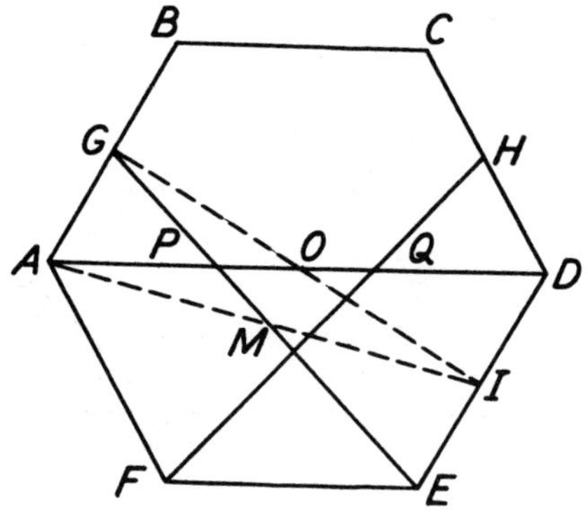

Figure 62

61. Let O_1, O_2, O_3 be the midpoints of $\overline{BC}$, $\overline{CA}$, $\overline{AB}$ respectively.

Since $\overline{O_2O_3} || \overline{BC} \perp \overline{CA} || \overline{EH}$, $\overline{O_2O_3} \perp \overline{EH}$ and therefore passes through the point of tangency T_4. Similarly for T_1, T_2, T_3. Since $\overline{EH}$ and $\overline{EF}$ are parallel to perpendicular lines $\overline{CA}$ and $\overline{CB}$, $\overline{EH} \perp \overline{EF}$. Similarly for other sides of $EFGH$, which is therefore a rectangle. If we prove $EF=EH$, the theorem is completed.

Since $EF=T_4T_2$ and $EH=T_1T_3$, we will prove $T_4T_2=T_1T_3$. Let $AC=b$, $BC=a$, $AB=c$.

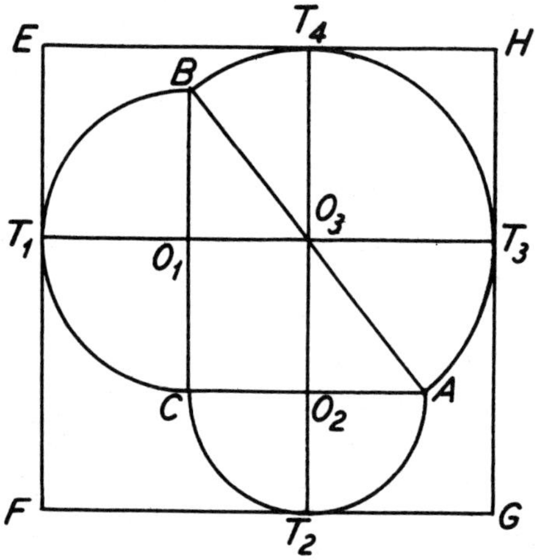

Figure 63

$$T_1T_3 = T_1O_1 + O_1O_3 + O_3T_3$$
$$= O_1C + \frac{1}{2}CA + O_3A$$
$$= \frac{1}{2}a + \frac{1}{2}b + \frac{1}{2}c = \frac{1}{2}(a+b+c).$$

Similarly T_2T_4 can be proven equal to $\frac{1}{2}(a+b+c)$. Therefore, $T_1T_3 = T_2T_4$ and $EFGH$ is a square.

62. Construct $\triangle BQC$ on side $\overline{BC}$ so that $m\angle QBC = m\angle ABP$, and $m\angle QCB = m\angle BAP$. This makes $\triangle BQC \cong \triangle ABP$, $BQ = BP(=3)$, and $QC = AP(=5)$.

Figure 64

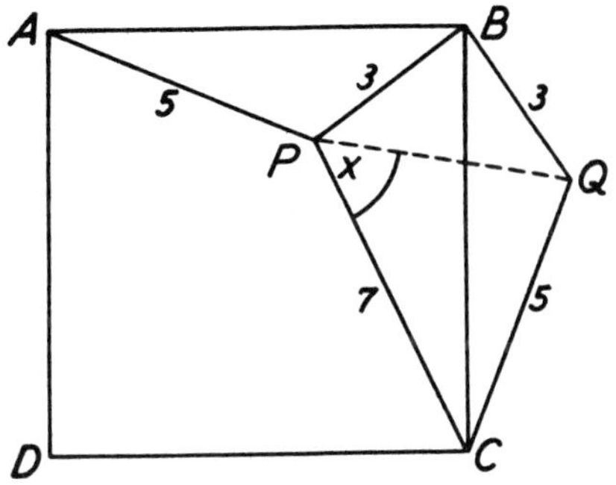

Since $m\angle QBC + m\angle PBC = m\angle ABP + m\angle PBC = 90°$, $\triangle PBQ$ is an isosceles right triangle, and $PQ = 3\sqrt{2}$. Apply the Law of Cosines in $\triangle PQC$:

$$\cos x = \frac{(PQ)^2 + (PC)^2 - (QC)^2}{2(PQ)\,(PC)} = \frac{1}{\sqrt{2}}.$$

Hence $m\angle x = 45°$, $m\angle BPC = 90°$, and $(BC)^2 = 3^2 + 7^2$. Thus $BC = \sqrt{58}$.

Notes:

The construction described above has the effect of rotating $\triangle ABP$ through $90°$, using B as a hinge. This type of transformation is a useful tool for solving certain problems. Following are two more examples of the use of such a transformation.

(i) If $\triangle ABC$ is equilateral and PA, PB, $PC = 4$, 3, 5, respectively, find the length of the side of the given triangle.

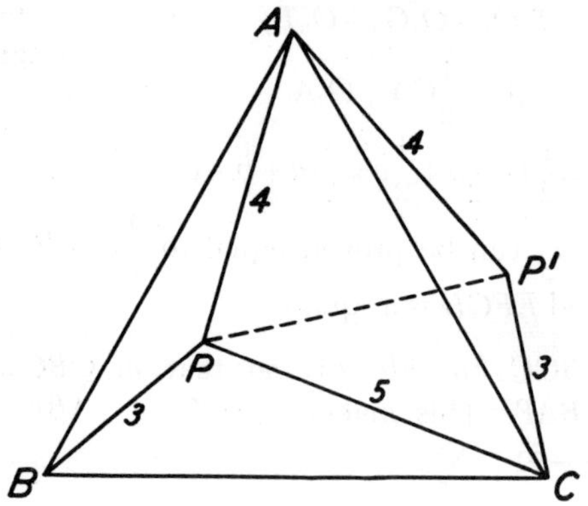

Figure 65

Rotate $\triangle ABP$ about vertex A so that $\overline{AB}$ coincides with $\overline{AC}$, as in Fig. 65. Draw $\overline{PP'}$ and prove: (*i*) $\triangle APP'$ is equilateral; (*ii*) $\triangle PP'C$ is a right triangle with sides: 3, 4, 5; (*iii*) $m\angle AP'C = 150°$. Now the length of side AC can be found by applying the Law of Cosines to $\triangle AP'C$.

(ii) For this problem refer to Fig. 56 in the notes to the solution of Problem 57.

Rotate $\triangle JEB$ about point B, so that $\overline{BE}$ coincides with $\overline{BA}$. Also rotate $\triangle FJB$ about point B, so that $\overline{BF}$ coincides with $\overline{BC}$ (see Fig. 66). Hence $EJ = AJ'$ and $FJ = CJ''$; $m\angle 1 = m\angle 3$ and $m\angle 2 = m\angle 4$; etc.

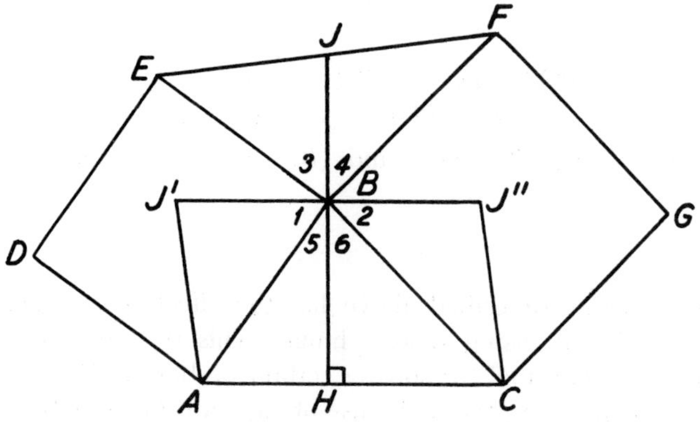

Figure 66

Since $m\angle EBA = 90°$ and $\angle JBH$ is a straight angle, it follows that $m\angle 3 + m\angle 5 = 90°$. But $m\angle 3 = m\angle 1$; therefore

$m\angle 1 + m\angle 5 = 90°$. In the same way it can be proven that $m\angle 2 + m\angle 6 = 90°$. Hence $\angle J'BJ''$ is a straight angle, and $\overline{J'BJ''}$ is a straight line parallel to $\overline{AC}$ (both$\perp BH$).

Now $m\angle J' = m\angle EJB$, and $m\angle J'' = m\angle FJB$. But $\angle EJB$ and $\angle FJB$ are supplementary; therefore $\angle J'$ is the supplement of $\angle J''$. Therefore $\overline{AJ'} \parallel \overline{CJ''}$, and $AJ'J''C$ is a parallelogram. Hence $AJ' = CJ''$, or $EJ = JF$.

(iii) Refer to the second solution of problem 42 (Fig. 27), in which the effect of the construction is to rotate $\triangle ABD$ into the position of $\triangle BD'C$.
Reference: [51].

63. Consider two right triangles RST and $R'S'T'$. If they have five pairs of equal parts and are not congruent, there must be three pairs of equal angles and two pairs of equal sides. Moreover, the equal sides cannot be correspondingly located with respect to pairs of equal angles. Let us suppose that $m\angle S = m\angle S' = 90°$, $m\angle R = m\angle R' = \theta$, and $m\angle T = m\angle T' = 90° - \theta$. We impose the conditions (1) $RS = R'T'$ and (2) $ST = R'S'$ in order to show that the resulting equation in θ has a solution θ_1 such that $0° < \theta_1 < 90°$. Let $RT = s$. Then in terms of s and θ, (1) becomes (3) $RS = s \cos \theta = R'T'$, and (2) becomes (4) $ST = s \sin \theta = R'S' = R'T' \cos \theta$. Combining (3) and (4) we obtain (5) $s \sin \theta = s \cos^2 \theta$, or $\cos^2 \theta - \sin \theta = 0$. Set $\cos^2 \theta = 1 - \sin^2 \theta$ in (5) and solve for $\sin \theta$ to obtain (6) $\sin \theta = \frac{-1+\sqrt{5}}{2}$. This value of $\sin \theta$ is positive and less than 1 so that (6) has a solution within the interval specified above. Call this solution θ_1. Then the lengths of the sides of $\triangle RST$ are $RT = s$, $ST = s \sin \theta_1$, $RS = s \cos \theta_1$, and the corresponding sides of $\triangle R'S'T'$ are $R'T' = s \cos \theta_1$, $S'T' = s \cos \theta_1 \sin \theta_1$, and $R'S' = s \cos^2 \theta_1$. These two triangles have two pairs of sides of equal length, ($RS = R'T'$ and $ST = R'S'$) and three pairs of angles of equal measure, and yet the triangles are not congruent. An approximate solution for θ is $38°10'$.

64. Draw $\overline{AC}$, $\overline{GH}$, $\overline{MN}$, $\overline{BD}$, each perpendicular to $\overleftrightarrow{PE}$, as in Fig. 67. Since $\triangle APC \cong \triangle GPH$, $AC = PH$ ($\angle APC$ corresponds to $\angle PGH$). Similarly, $BD = HE$. Therefore $AC + BD = PH + HE = PE$. Now $\overline{MN}$ is a line joining the midpoints of the legs of trapezoid $ACDB$, because the three parallel lines $\overline{AC}$, $\overline{MN}$, and $\overline{BD}$ cut off equal segments on $\overline{AB}$, and therefore cut off equal segments on $\overline{CD}$. Therefore $MN = \frac{1}{2}(AC + BD) = \frac{1}{2}PE$. And since $CP = GH = ED$, N is the midpoint of $\overline{PE}$. (Compare with the solution of Problem 20.) Thus M can be located by constructing the perpendicular bisector of $\overline{PE}$ (on the side of the missing gallows) and laying off $NM = \frac{1}{2}\ PE$. This construction is independent of the position of G.

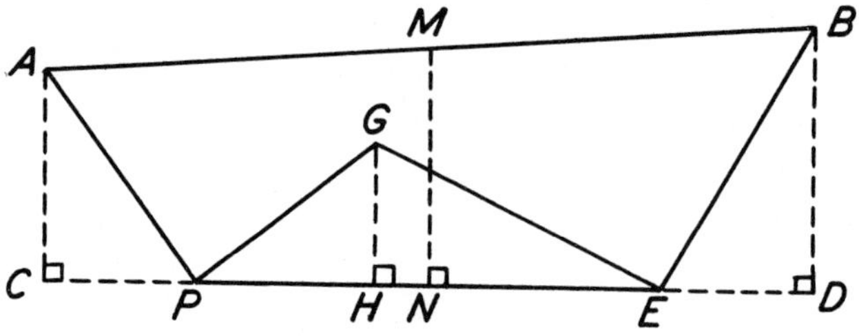

Figure 67

Note:

If G were (a) south of the line $\overleftrightarrow{PE}$, (b) on segment $\overline{PE}$, (c) on $\overleftrightarrow{PE}$ prolonged, the problem would still be valid.

A second solution, requiring a knowledge of the properties and graphing of complex numbers will be found in the book: *One, Two, Three . . . Infinity,* by George Gamow. (Mentor Books, 1953. pp. 44-47.)

65. **First Solution**

Let the two lines be designated as l_1 and l_2, and let D be the given point. To obtain a line through D which will pass through the inaccessible intersection, A, of l_1 and l_2, proceed as follows: draw $\overleftrightarrow{DF}$ perpendicular to l_2 intersecting l_2 at F and l_1 at B. Draw $\overleftrightarrow{DE}$ perpendicular to l_1 intersecting l_1 at E and l_2 at C; then draw $\overline{BC}$. Now draw $\overrightarrow{DT}$ perpendicular to $\overline{BC}$. $\overrightarrow{DT}$ will pass through the inaccessible intersection, A, of $\overleftrightarrow{BE}$ (l_1) and $\overleftrightarrow{CF}$ (l_2) because the three altitudes of $\triangle BCA$ must be concurrent.

Note:

This solution is valid whether D is taken between the two lines, or outside of them. If D is taken on one of the given lines, the solution is trivial.

Second Solution

This solution can be effected with straightedge alone. Construct any line through D intersecting l_1 and l_2 at B and C respectively. Determine H, the harmonic conjugate of D with respect to B and C (Appendix D). Through H construct any line l_3 intersecting l_1 and l_2 at J and K respectively. If $\overline{BK}$ and $\overline{CJ}$ intersect at point L, $\overleftrightarrow{DL}$ is the required line. See Fig. 68.

Figure 68

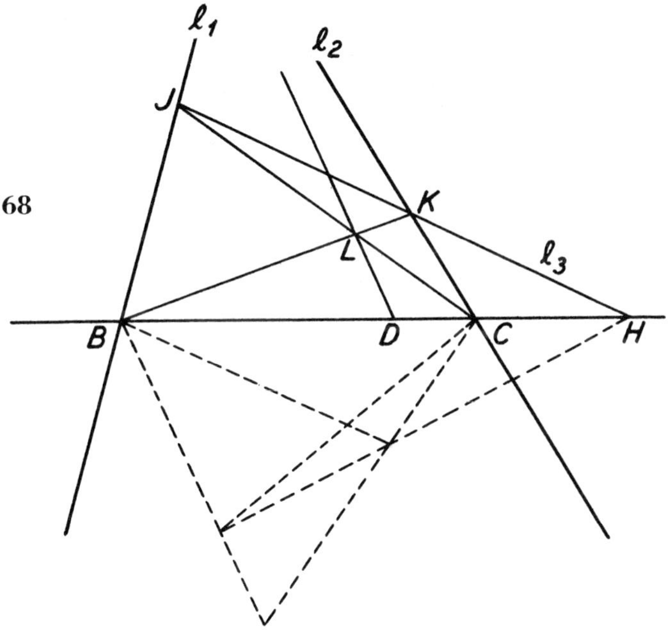

Third Solution

If D is the midpoint of $\overline{BC}$, then its harmonic conjugate is a point at infinity, and $\overline{EF}$ will be parallel to $\overline{BC}$. This suggests the following procedure:

Draw a line through D intersecting l_1 and l_2 at B and C respectively such that $BD=DC$. (Through D draw any line intersecting l_1 at P. Extend $\overrightarrow{PD}$ to Q so that $PD=QD$. Through Q draw a line parallel to l_1 intersecting l_2 at C. $\overrightarrow{CD}$ will intersect l_1 at B such that $BD=DC$.) Draw any line parallel to $\overline{BC}$ intersecting l_1 and l_2 at

F and E respectively. If $\overline{BE}$ and $\overline{CF}$ intersect at G, $\overrightarrow{DG}$ is the required line.

Note:

In all these constructions it is assumed that all points of l_1 and l_2 are accessible except their intersection.

66.

First Solution

Draw the circle determined by R, Q, D, and let it intersect $\overline{AC}$ again at I. Draw the circle determined by B, P, Q, and let it intersect $\overline{AC}$ again at H. We have $AB \cdot AP = AH \cdot AQ$, and $AD \cdot AR = AI \cdot AQ$. Adding these, we obtain $AB \cdot AP + AD \cdot AR = AQ(AH + AI)$. But $\triangle CID \cong \triangle BHA$, because $CD = AB$, $m\angle ICD = m\angle BAH$, and $m\angle CID$ $(= m\angle QRD = m\angle APQ) = m\angle BHA$, (because of successive inscribed quadrilaterals. $\angle CID$ and $\angle QRD$ are both supplements of $\angle QID$, and are therefore equal. Apply this to the successive quadrilaterals: $DRQH$, $PARQ$, and $BHQP$.) Therefore $AH = IC$, and $AH + AI = AI + IC = AC$. Finally, $AB \cdot AP + AD \cdot AR = AC \cdot AQ$.

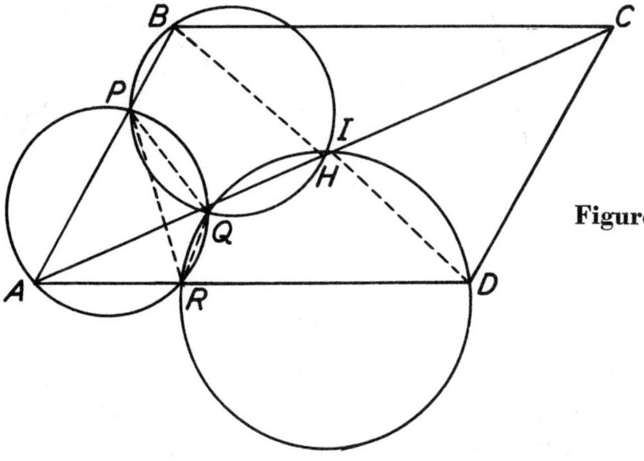

Figure 69

Second Solution

(refer to Fig. 69)

Apply Ptolemy's Theorem (Appendix F) to quadilateral *APQR:*
(1) $AP \cdot QR + AR \cdot PQ = AQ \cdot PR$.

But $\triangle PQR \sim \triangle ACD$, because $m\angle QPR = m\angle QAR$ and $m\angle PRQ$ $(= m\angle PAQ) = m\angle ACD$. Hence $\frac{PQ}{AD} = \frac{QR}{CD} = \frac{PR}{AC} = t$, where t is chosen arbitrarily. Therefore $PQ = t(AD)$, $QR = t(CD)$, and $PR = t(AC)$. Substitute these values in (1), divide by t, and replace CD by its equal AB; then the required result is obtained.

67. Draw $\overline{OC}$, $\overline{OP}$, $\overline{OM}$, $\overline{OD}$, and $\overline{OQ}$, where O is the center of the circle. Therefore C, P, M, O are concyclic (Appendix E) and $m\angle 1 = m\angle 2$. Also M, O, Q, D are concyclic, and $m\angle OQD = m\angle 2$ (both are supplements of $\angle OMD$). Therefore $m\angle 1 = m\angle OQD$, $\triangle OCP \cong \triangle ODQ$, (SAA), and $CP = QD$, one of the required results.

Figure 70

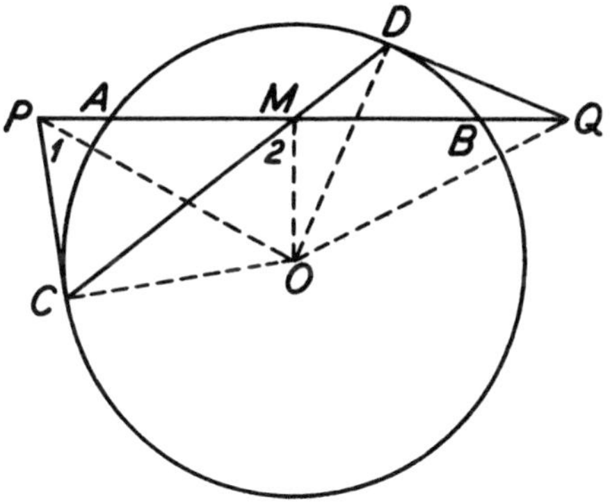

Now $(PC)^2 = (PA)(PB)$, and $(QD)^2 = (QB)(QA)$. Therefore $PA \cdot PB = QB \cdot QA$, or $PA(PA + AB) = QB(QB + AB)$. Thus $(PA)^2 + PA \cdot AB = (QB)^2 + QB \cdot AB$. Then $(PA)^2 - (QB)^2 + PA \cdot AB - QB \cdot AB = 0$, and $(PA + QB)(PA - QB) + AB(PA - QB) = 0$. The common factor is $PA - QB$; hence $(PA - QB)(PA + QB + AB) = 0$. Only the first factor can equal zero; therefore $PA = QB$.

There are at least two other methods of reaching this result, beginning with $(PA)^2 + PA \cdot AB = (QB)^2 + QB \cdot AB$.

(i) $\frac{PA}{QB + AB} = \frac{QB}{PA + AB}$. Therefore $\frac{PA}{PA + QB + AB} = \frac{PA}{PA + QB + AB}$ (Appendix A) and $PA = QB$.

(ii) Complete the square on each side:

$$(PA)^2 + PA \cdot AB + \frac{(AB)^2}{4} = (QB)^2 + QB \cdot AB + \frac{(AB)^2}{4}.$$

Therefore $\left(PA + \frac{AB}{2}\right)^2 = \left(QB + \frac{AB}{2}\right)^2$, and $PA = QB$.

68. Draw $\overline{ED}$. In $\triangle AED$, $\frac{AD}{AE} = \frac{1}{2}$, and $m\angle A = 60°$. This is possible only if $\triangle AED$ is a right triangle with right angle at D.

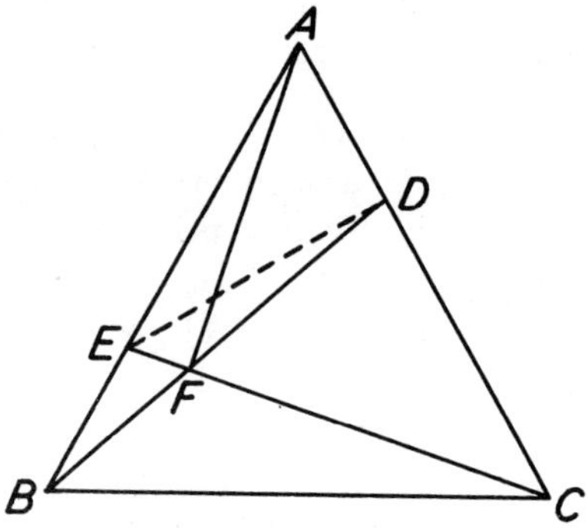

Figure 71

Now $\triangle CEB \cong \triangle ABD$, (SAS); therefore $m\angle BEC = m\angle ADF$, and $\angle AEF$ is the supplement of $\angle ADF$. Hence A, E, F, D are concyclic (Appendix E) and $m\angle AFE = m\angle ADE$. But $m\angle ADE = 90°$; therefore $\overline{AF} \perp \overline{EC}$.

III

PROBLEMS WITH AN ALGEBRAIC BACKGROUND

69. The following table gives the distances run, when Bob and Jim each complete two miles.

Bob	Jim	Henry
2	$\frac{19}{10}$	x
	2	$\frac{9}{5}$

Therefore $\frac{19}{10}:2=x:\frac{9}{5}$, and $x=1.71$. Therefore Bob beats Henry by .29 miles.

70. The following table gives the distances run by A, B, and C.

A	B	C
d	$d-p$	$d-r$
	d	$d-q$

Therefore $(d-p):d=(d-r):(d-q)$. Solving for d, we obtain $d=\frac{pq}{p+q-r}$.

Note:

For the problem to be valid, $p+q$ must be greater than r. This may be confirmed by intuitive consideration, and also by the fact that the denominator of the answer must be positive.

71. **First Solution**

Let the equation of the straight line be $bx+ay=ab$. Then the coordinates of C are (O, b) and of D are (a, O). Let the coordinates of A be (x_1, y_1), and of B, (x_2, y_2).

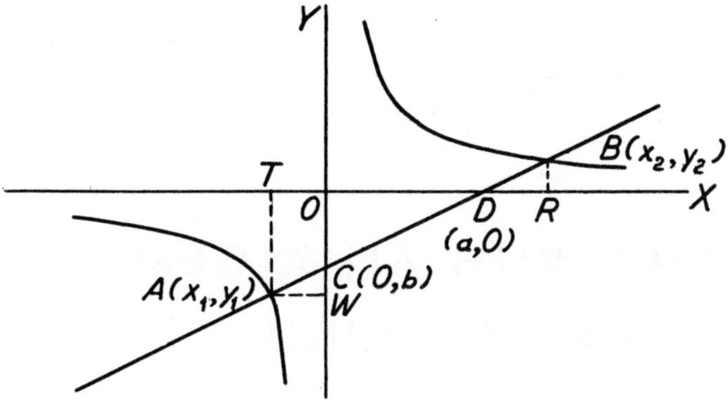

Figure 72

To prove that $AC=BD$, then, by the distance formulas we must prove:

(i) $x_1^2+(y_1-b)^2=(x_2-a)^2+y_2^2$; or

(ii) $x_1^2-(x_2-a)^2=y_2^2-(y_1-b)^2$; or factoring,

(iii) $(x_1+x_2-a)\,(x_1-x_2+a)=(y_2+y_1-b)\,(y_2-y_1+b)$.

Multiply both members of $bx+ay=ab$ by x:

(iv) $bx^2+axy=abx$; or, since $xy=k$,

(v) $bx^2-abx+ak=0$.

Now the roots of equation (**v**) are x_1 and x_2, the abscissas of A and B. By the formula for the sum of the roots:

(vi) $x_1+x_2=a$. Similarly it can be shown that:

(vii) $y_1+y_2=b$.

Substituting (**vi**) and (**vii**) into (**iii**) verifies that (**iii**) is true. Then, since the steps are reversible, $AC=BD$.

Second Solution

From Fig. 72, $AW=x_1=TO$, $AT=y_1$, $x_2=OR$, and $RB=y_2$. Since A and B are on the curve, their coordinates satisfy the equation: $xy=k$. Therefore $x_1y_1=x_2y_2$ $(=k)$, or $TO \cdot AT=OR \cdot RB$, whence $\frac{RB}{AT}=\frac{TO}{OR}$. But $\triangle BDR \sim \triangle TDA$; therefore $\frac{RB}{AT}=\frac{DR}{TD}$. These two proportions give: $\frac{DR}{TD}=\frac{TO}{OR}$, or $\frac{DR+TD}{TD}=\frac{TO+OR}{OR}$ (Appendix A). Since the numerators are equal $(=TR)$, it follows that

$TD = OR$. Subtracting OD from each, we obtain $TO = DR$, or $AW = DR$. Similarly, it can be proven that $CW = BR$. Hence

$$(AW)^2 + (CW)^2 = (DR)^2 + (BR)^2,$$

from which the required result follows.

Note:

(i) If (x_1, y_1) and (x_2, y_2) are interchanged, the result is still valid, for if $AD = BC$, then $AC = BD$.

(ii) If a straight line intersects the hyperbolas $xy = k_1$ and $xy = k_2$ then the segments cut off by the hyperbolas are equal. This is readily proven by applying the results of the problem proven above.

72. Let x be the speed of the current. Then going downstream, the boat's rate is $24 + x$, and its time is $\frac{60}{24+x}$. Going upstream, the boat's rate is $24 - x$, and its time is $\frac{60}{24-x}$. Therefore its average rate (see solution to Problem 9) is $\frac{120}{\frac{60}{24+x} + \frac{60}{24-x}}$, which is given as 18. The solution to this equation is $x = 12$.

As shown in the reference mentioned, the distance (in this case 120 miles) is not needed.

73. Let M be the midpoint of $\overline{AB}$, and let w = rate of walking, r = rate of rowing in still water, and c = rate of the current. Each term in the following equations represents the time taken to travel a designated distance:

	Row: B to M		Walk: M to A		Total Time
(1)	$\frac{12}{r+c}$	+	$\frac{12}{w}$	=	5
	Walk: A to M		**Row: M to B**		
(2)	$\frac{12}{w}$	+	$\frac{12}{r-c}$	=	7
	Walk		**Row**		
(3)	$\frac{12}{w}$	+	$\frac{12}{r}$	=	$\frac{17}{3}$

Solve for $\frac{12}{w}$ in each:

(4) $\frac{12}{w} = 5 - \frac{12}{r+c} = 7 - \frac{12}{r-c} = \frac{17}{3} - \frac{12}{r}$; therefore

(5) $\dfrac{12}{w}=\dfrac{5r+5c-12}{r+c}=\dfrac{7r-7c-12}{r-c}=\dfrac{17r-36}{3r}.$

Use the following theorem in proportions: *If* $\dfrac{a}{b}=\dfrac{c}{d}$, *then either fraction is equal to* $\dfrac{a+c}{b+d}$, *i.e. equals (numerator + numerator) over (denominator + denominator).*

Apply this to the second and third fractions in (5).

(6) $\dfrac{5+5c-12}{r+c}=\dfrac{7r-7c-12}{r-c}=\dfrac{12r-2c-24}{2r}$

$\left[=\dfrac{17r-36}{3r}\text{, the last fraction in (5)}\right].$

The last two fractions yield $r=3c$. Substituting in (5), gives: $r=\dfrac{9}{2}, c=\dfrac{3}{2}, w=4.$

Let $\dfrac{a}{b}=\dfrac{c}{d}=t$; then $a=bt$, $c=dt$. Now $\dfrac{a+c}{b+d}=\dfrac{t(b+d)}{b+d}=t$, which equals each of the original fractions.

A more general theorem states that if $\dfrac{a_1}{b_1}=\dfrac{a_2}{b_2}=\ldots=\dfrac{a_n}{b_n}$, then each of these fractions equals $\dfrac{p_1a_1+p_2a_2+\ldots+p_na_n}{p_1b_1+p_2a_2+\ldots+p_na_n}$, for any p_i's, provided that the denominator does not become zero. Reference: [18], Chapters I, II.

74. When the hikers set out, let $x=$ the distance traveled uphill, $y=$ the distance traveled on level land, and $z=$ the distance traveled downhill. The total time for the outward trip is $\dfrac{x}{3}+\dfrac{y}{4}+\dfrac{z}{6}$.

Returning, the distances uphill and downhill reverse. Therefore the total time for the return trip is $\dfrac{x}{6}+\dfrac{y}{4}+\dfrac{z}{3}$. The total time for the entire trip becomes $\dfrac{x+y+z}{2}=6$, or $x+y+z=12$, for the distance in one direction. Hence the hikers walked 24 miles.

Note:

This problem is based on one by Lewis Carroll (pseudonym of Charles Lutwidge Dodgson: 1832-1898), famous author of *Alice in Wonderland, Through the Looking Glass, Bruno and Sylvie,* numerous poems, puzzles, and an enormous variety of miscellaneous articles. He also wrote a number of books on mathematics. The problem referred to occurs as "Knot I" in

A Tangled Tale. It is given verbatim, together with Lewis Carroll's solution.

Problem

Two travelers spend from 3 o'clock till 9 walking along a level road, up a hill, and home again. Their pace on the level road is 4 miles an hour, up hill 3, and down hill 6. Find distance walked and also determine (within half an hour) the time of reaching the top of the hill.

Solution

A level mile takes $\frac{1}{4}$ of an hour, up hill $\frac{1}{3}$, down hill $\frac{1}{6}$. Hence to go and return over the same mile, whether on the level or on the hill-side, takes $\frac{1}{2}$ an hour. Hence, in 6 hours they went 12 miles out and 12 back. If the 12 miles out had been nearly all level, they would have taken a little over 3 hours; if nearly all up hill, a little under 4. Hence $3\frac{1}{2}$ hours must be within $\frac{1}{2}$ an hour of the time taken in reaching the peak; thus, as they started at 3, they got there within $\frac{1}{2}$ hour of half past six. Reference: [3].

75. Let x = amount of grass one cow can eat in one week, y = amount of grass on one acre, and z = amount of grass which grows on one acre in one week. Since two cows in four weeks can eat all the grass on two acres and all the grass which grows on these two acres during the four weeks, we have $8x=2y+8z$. Also we know that three cows in two weeks can eat all the grass on two acres and all the grass that grows on these two acres during the two weeks, whence $6x=2y+4z$. Now if n represents the minimum number of cows required to eat in six weeks all the grass on six acres and all the grass that grows on these six acres during the six weeks, then $6nx=6y+36z$. Eliminating x and y from our three equations (solve for x and y in the first two equations in terms of z, and substitute in the third), we obtain $12nz=24z+36z=60z$. Hence, $n=5$.

76. Let x represent the length in inches of the longer candle, and let r denote its rate of burning expressed in inches per hour. Then $x-1$ represents the length of the shorter candle; let s denote its rate of burning. At 8:30, the longer candle has burned for 4 hours and the shorter candle for $\frac{5}{2}$ hours. Since they are then of the same length,

$$x-4r=(x-1)-\frac{5s}{2}.$$

Also the longer candle is consumed in six hours and the shorter candle is consumed in four hours. This yields the equations

$$6r=x,$$

and

$$4s=x-1.$$

The solution of these three equations yields $r=1.5$, $s=2$, and $x=9$. Therefore the longer candle is 9 inches long, and the shorter is 8 inches long.

77. Let D represent the number of hours required for the chauffeur to drive from the commuter's home to the boat dock, and let W represent the number of hours the commuter had been walking. Then $2D$ is the number of hours usually required for the round trip. In this case the round trip was completed one half hour earlier than usual so that the chauffeur drove for $\left(2D-\frac{1}{2}\right)$ hours. The commuter arrived at the dock one hour earlier than usual, and he had been walking for W hours when he met the car. Therefore, the chauffeur met the commuter $(1-W)$ hours earlier than usual, and the chauffeur had been driving for $D-(1-W)$ hours when this happened. The return trip took the same length of time, so the chauffeur drove for a total of $2[D-(1-W)]$ hours. Evidently $2D-\frac{1}{2}=2[D-(1-W)]$ from which we obtain $W=\frac{3}{4}$. The commuter had been walking three quarters of an hour.

78. **First Method**

If $x^2-6x+8>0$, then $(x-4)(x-2)>0$.
This will be true only if both factors have the same sign. The two possibilities are therefore:

(1) $x-4>0$ and $x-2>0$ simultaneously, which is satisfied by $x>4$, or

(2) $x-4<0$ and $x-2<0$ simultaneously, which is satisfied by $x<2$.

Second Method

An examination of the graph of $y=x^2-6x+8$ will reveal that y is positive only when $x<2$, or when $x>4$.

79. If a and b are the two roots of the given equation $x^2-px+q=0$, then $a+b=p$ and $ab=q$. If $a+\frac{1}{b}$ and $b+\frac{1}{a}$ are the roots of the desired equation, then the sum of its roots is $a+b+\frac{1}{a}+\frac{1}{b}$, and the product of the roots is $ab+\frac{1}{ab}+2$. Hence the required equation is $x^2-\left(p+\frac{p}{q}\right)x+\left(q+\frac{1}{q}+2\right)=0.$

Note:

See solution to Problem 11.

80. Let x be the number of gallons of water removed. After removing this water, the tank then contains $250-x$ gallons of water. Now x gallons of alcohol are added to the tank. The tank contains x gallons of alcohol and $250-x$ gallons of water. Next, x gallons of this mixture are removed. That is, $x\left(\frac{x}{250}\right)$ gallons of alcohol are removed, since the mixture is $\frac{x}{250}$ alcohol. Then $x-\frac{x^2}{250}$ is the number of gallons of alcohol remaining. Since x gallons of alcohol are then added, the final mixture contains $250-\left(x+x-\frac{x^2}{250}\right)$ gallons of water. Therefore,

$$\frac{250-\left(x+x-\frac{x^2}{250}\right)}{x+x-\frac{x^2}{250}}=\frac{16}{9}.$$

Solving, $x=50=$number of gallons of water removed the first time. Since 50 gallons of alcohol are then added, the tank contains 50 gallons of alcohol and 200 gallons of water. That is, $\frac{1}{5}$ is alcohol and $\frac{4}{5}$ is water. Then 50 gallons of this mixture are removed, so 40 gallons of water are removed the second time. The final mixture contains 90 gallons of alcohol and 160 gallons of water.

81. Let the length of each candle be d inches. Let $x=$the number of hours required for one to be three times the length of the other.

Then $d-\frac{dx}{5}=3\left(d-\frac{dx}{4}\right)$ Divide through by d and solve for x.
$x=\frac{40}{11}$ hours.

Note:

The time required is independent of the initial lengths of the candles.

82. Let the first term of the geometric progression be $\frac{x}{r}$, and let the common ratio be r. Then

$$x\left(\frac{1}{r}+1+r\right)=21, \text{ and}$$

$$\frac{1}{x}\left(\frac{1}{r}+1+r\right)=\frac{7}{12}$$

By division, we obtain: $x^2=36$, and $x=\pm 6$.

Substituting $x=6$ in the first equation we obtain $r=2$ or $\frac{1}{2}$.
Thus the three required integers are: 3, 6, 12 or 12, 6, 3.
Letting $x=-6$ leads to irrational values of r.

83. If the two equations are multiplied there results: $(y-x)^2=9$. Therefore $y-x=+3$ or -3.

If we substitute $y-x=+3$ in $\frac{y-x}{xy}=\frac{1}{36}$, we obtain $xy=108$.
Now $(y-x)^2+4xy=9+432$.
Hence $(y+x)^2=441$, and $y+x=+21$ or -21.
Combining these with $y-x=3$, and recalling that the product of x and y is positive, we obtain the answers $(x, y)=(9, 12)$ or $(-12,-9)$.

If we let $y-x=-3$ and use the same technique, we obtain $y+x=+i\sqrt{423}$ or $-i\sqrt{423}$. The rest is left to the reader.

84. Let $x+y=u$, and $xy=v$. The given equations then become:

$$u^2-2v+u=8$$
$$v+u=5$$

These are easily solved to obtain $(u, v)=(3, 2)$ or $(-6, 11)$. Replacing u and v by $x+y$ and xy respectively, we obtain in the first case: $x+y=3$, and $xy=2$. These can be solved in a straightforward manner; but a useful device is to consider x and y to be the roots of some quadratic equation. Since the sum of these roots is 3 and their product is 2, that quadratic equation might be

$Z^2-3Z+2=0$, whose roots are 2 and 1. Hence $(x, y)=(2, 1)$ or $(1, 2)$. Applying the same method to the case in which $(u, v)=(-6, 11)$, we obtain the roots $(x, y)=(-3+i\sqrt{2}, -3-i\sqrt{2})$ or $(-3-i\sqrt{2}, -3+i\sqrt{2})$.

85. From the given data $x^2+y^2=\frac{x}{a}=\frac{y}{b}=t$ (where t is chosen arbitrarily). Therefore $x=at$ and $y=bt$. Substitute in the first given equation: $a=\frac{at}{t^2(a^2+b^2)}$. Hence $t=\frac{1}{a^2+b^2}$, and t also equals $\frac{x}{a}=\frac{y}{b}$. Therefore $x=\frac{a}{a^2+b^2}$; $y=\frac{b}{a^2+b^2}$.

Problem

Solve the same problem beginning with

$$a=\frac{x}{px^2+qy^2};\ b=\frac{y}{px^2+qy^2}.$$

Note:

These are special cases of a more general problem that the reader may wish to investigate:

Find $f(x, y)$ and $g(x, y)$ such that if $f(x, y)=a$ and $g(x, y)=b$, then $f(a, b)=x$ and $g(a, b)=y$.

86. $N=\dfrac{1-\left(\frac{1}{2}\right)^n}{1-\frac{1}{2}}=2-\frac{1}{2^{n-1}}$; therefore $2-N=\frac{1}{2^{n-1}}$ and $\frac{1}{2^{n-1}}<.0001$. It is easy to verify by trial, or by logarithms, that the smallest integral value of n is 15.

Note:

This problem presents a situation basic to the topic of limits in calculus.

87.

First Solution

$1°$ on the A scale $=\frac{16}{24}$ or $\left(\frac{2}{3}\right)°$ on the B scale. $1°$ on the B scale $=\frac{27}{12}$ or $\left(\frac{9}{4}\right)°$ on the C scale. Thus $1°$ on the A scale $=\left(\frac{2}{3}\right)\left(\frac{9}{4}\right)=\left(\frac{3}{2}\right)°$ on the C scale. Therefore a drop of $18°$ on the A scale $=-18\left(\frac{3}{2}\right)$ or a drop of $27°$ on the C scale.

Second Solution

If $\triangle A$, $\triangle B$, $\triangle C$ represent corresponding changes of temperature on the A, B, C, scales, then $\frac{\triangle B}{\triangle A}=\frac{16}{24}=\frac{2}{3}$; $\frac{\triangle C}{\triangle B}=\frac{27}{12}=\frac{9}{4}$. Therefore $\frac{\triangle C}{\triangle A}=\left(\frac{\triangle B}{\triangle A}\right)\cdot\left(\frac{\triangle C}{\triangle B}\right)=\left(\frac{2}{3}\right)\left(\frac{9}{4}\right)=\frac{3}{2}$. Thus if $\triangle A=-18°$, $\triangle C=-27°$.

88. Since y is a monomial, $y=ax^b$ and $ay^b=3x^2$. Thus $a(ax^b)^b=3x^2$, and $(a^{b+1})(x^{b^2})=3x^2$. By comparison of the two members we see: $b^2=2$, and $b=\pm\sqrt{2}$. Hence $a^{\pm\sqrt{2}+1}=3$, or $a=3^{\pm\sqrt{2}-1}$. The required relationship may be written: $y=\frac{(3x)^{\pm\sqrt{2}}}{3}$.

Note:

The problem might have been stated as follows:

If $f[f(x)]=3x^2$, find $f(x)$.

89. The left members of the equations factor as follows:

(1) $(2x+y-4)(x-y)=0$;

(2) $(2x+y-4)(3x-y-2)=0$.

Thus, each graph consists of two intersecting lines. The four points $(-1, 6)$, $(3, -2)$, $(0, 4)$, and $(2, 0)$, lying on the straight line $2x+y-4=0$, satisfy both equations, and $(1, 1)$, being the intersection of $x-y=0$ and $3x-y-2=0$, also satisfies both equations. Furthermore, any point on $2x+y-4=0$ will satisfy equations (1) and (2). Thus, the original statement is valid only for quadratics that have no common factor.

Note:

Problems 88 and 89 were contributed by Walter Penney, U.S. Government, Department of Defense.

90. If the distance from C to D is x, the given data translates into the equation:

$$\frac{x+14}{2}=7+\frac{1}{3}(x+7)-\frac{1}{4}\left[\frac{1}{3}(x+7)+1\right].$$

Solving this we obtain: $x=6$.

91. Let the towns P, Q, R, S correspond to towns A, B, C, D of Problem 90. Then following the first direction brings us to $\frac{x_1+x_2}{2}$.

After the second direction we are at $\frac{x_1+x_2}{2}+\frac{1}{3}\left[x_3-\left(\frac{x_1+x_2}{2}\right)\right]$
$=\frac{x_1+x_2+x_3}{3}$.

After the final direction we are at:

$$\frac{x_1+x_2+x_3}{3}+\frac{1}{4}\left[x_4-\frac{x_1+x_2+x_3}{3}\right]=\frac{1}{4}(x_1+x_2+x_3+x_4).$$

Since the names of the towns were illegible, the directions could be followed in 24 different ways corresponding to the 24 permutations of A, B, C, D. Each permutation has the effect of permuting the x's in the answer. It is plain that the answer is not affected by such changes. Therefore the treasure will be found at

$$\frac{x_1+x_2+x_3+x_4}{4}.$$

Notes:

(i) The problem may be generalized for n towns, in which case the directions would be expanded so that the fourth would read: "Then go $\frac{1}{5}$ of the way to ___; the fifth: "Then go $\frac{1}{6}$ of the way to ___; etc. For n towns the treasure would be located at $\frac{x_1+x_2+\ldots+x_n}{n}$. This can be proven by mathematical induction, making use of the identity:

$$\frac{x_1+x_2+\ldots+x_n}{n}+\frac{1}{n+1}\left[x_{n+1}-\frac{x_1+x_2+\ldots+x_n}{n}\right]=\frac{x_1+x_2+\ldots+x_{n+1}}{n+1}.$$

(ii) A further generalization can be made for any set of points in a plane (or even in space). For example:

Towns exist at $P_1:(x_1, y_1)$, $P_2:(x_2, y_2)$, ... $P_n(x_n, y_n)$, and a treasure is located as follows:

(1) Start at any town and go $\frac{1}{2}$ of the way to any other town.

(2) Then go $\frac{1}{3}$ of the way to any other town.

. . . .

(n) Then go $\frac{1}{n}$ of the way to the remaining town.

The coordinates of the final point will be:

$$X=\frac{x_1+x_2+\ .\ .\ .\ +x_n}{n};\ Y=\frac{y_1+y_2+\ .\ .\ .\ +y_n}{n}.$$

To prove this, drop perpendiculars from $P_1, P_2, \ldots P_n$ to the x-axis. Their abscissas will be: $x_1, x_2, \ldots x_n$, and the result of following the directions, according to note (i), will be $\frac{x_1+x_2+\ .\ .\ .\ +x_n}{n}$. Then drop perpendiculars from $P_1, P_2, \ldots P_n$ to the y-axis to obtain $\frac{y_1+y_2+\ .\ .\ .\ +y_n}{n}$.

(iii) The point at which the treasure is located is called the CENTROID with respect to the given set of points. If equal weights are placed at each of these points, the CENTER OF GRAVITY of the system is located at the centroid. This is the point at which the weight of the system may be considered to be concentrated.

(iv) Apply the theorem to the vertices of $\triangle ABC$. For example: (1) Start at A and go $\frac{1}{2}$ of the way to B; (2) then go $\frac{1}{3}$ of the way to C. Each of the six ways of doing this end up at the same point G. This constitutes a proof that the medians of a triangle are concurrent.

(v) A great many problems can de devised based on this idea. For example:

Let E be the midpoint of side $\overline{AD}$ of square $ABCD$. Take F on $\overline{EB}$ so that $EF=\frac{1}{3}EB$. Prove: (a) F, C, and the intersection G, of the diagonals of the square are collinear; (b) $FG=\frac{1}{4}FC$.

92. If $f(x)=x^2+ax+bc=0$ and $g(x)=x^2+bx+ca=0$ have a common root, then $f(x)$ and $g(x)$ have a common factor. They would have this factor in common with $f(x)-g(x)$, viz. $h(x)=(a-b)x+c(b-a)$. Hence the common root may be obtained by solving $h(x)=0$; viz. $x=\frac{c(a-b)}{a-b}$. Now $a\neq b$, for if they were equal, $f(x)$ and $g(x)$ would be identical and have more than one common factor, contrary to the hypothesis. Hence $x=c$ is the common root.

Since the product of the roots of $f(x)=0$ is bc, its other root is b. Similarly, the other root of $g(x)$ is a. Also, since the sum of the roots of $f(x)=0$ is $-a$, $b+c=-a$, or $a+b=-c$. This confirms that $x^2+cx+ab=0$ has a and b (the *other* roots) as its roots.

93. If (x, y) is any point on the circle described, then its distance from B is equal to the length of $\overline{AB}$. Therefore, by the distance formula:

$$\left(x+\frac{p}{2}\right)^2+\left(y-\frac{q+1}{2}\right)^2=\left(\frac{-p}{2}\right)^2+\left(\frac{q-1}{2}\right)^2.$$

If, in this equation, y is set equal to zero, the resulting equation will give the abscissas of the intersection of the circle with the x-axis. If the indicated operations are performed, the resulting equation is $x^2+px+q=0$, which is the given equation.

IV

INFERENTIAL PROBLEMS

94. Problems of this type are usually solved by the method of elimination. A convenient device for keeping track of one's steps is to record them in a tabular form, such as the following.

	Mr. B	Mr. C	Miss G	Mrs. J	Miss L	Mr. O
Manager	x		x		x	
Assistant Manager						
Cashier	x		x		x	
Teller						
Treatment Clerk						
Stenographer	x		x		x	

As a correspondence between position and name is eliminated, an **x** is placed in the appropriate box. The crosses shown result from the first step below.

(a) Since the Manager, Stenographer, and Cashier are married, these positions cannot be filled by Mr. Brown, Miss Gordon, or Miss Leonard.

(b) The Assistant Manager, and the Cashier are males. Therefore these positions cannot be filled by Miss Gordon, Mrs. Johnson, or Miss Leonard.

(c) The Teller is female, and therefore cannot be Mr. Brown, Mr. Cowan, or Mr. O'Shaunessey.

(d) From statement (f) in the problem Mr. O'Shaunessey is not the Manager.

(e) The Manager is a grandfather, and therefore is not Mr. Cowan.

At this point all but Mrs. Johnson have been eliminated as Manager. Place a circle in the box indicating this, and crosses indicating that Mrs. Johnson is not the Teller, Treatment Clerk, or Stenographer.

The reader should have no trouble completing the table from additional clues. It will be found that Mr. Cowan is the Cashier. The other positions are: Assistant Manager—Mr. Brown, Teller—Miss Leonard, Treatment Clerk—Miss Gordon, and Stenographer—Mr. O'Shaunessey.

95. According to the conditions of the problem, the visitor knows that there are fewer than eighteen children playing in the garden, that these children belong to four different families, and that no two families are represented by the same number of children. The visitor also knows that the product of these four numbers is the house number which he saw when he came in. With this information he thinks he may be able to find the number of children in each of the four families. Of course, we do not know the house number, but we can understand the method of attack which would be used by one who does know it. The visitor evidently intends to express the house number as the product of four factors, no two of which are equal. If this can be done in only one way the problem is solved. For example, if the house number happened to be 210, the four factors are 2, 3, 5, and 7 and these numbers tell us how many children there are in each of the four families. From the fact that our visitor needs more information, we conclude that the house number can be expressed as the product of four factors in more than one way. If we write down all possible products of four different numbers whose sum is less than 18, we find that there are a number of cases where the same product is obtained from two different sets of factors. For example, $96 = 1 \cdot 3 \cdot 4 \cdot 8 = 1 \cdot 2 \cdot 6 \cdot 8$, $80 = 1 \cdot 2 \cdot 4 \cdot 10 = 1 \cdot 2 \cdot 5 \cdot 8$, etc. There is only one case, however, where the visitor's question, "Does your cousin's family consist of a single child?," would be significant. This is in case the house number was 120, because $120 = 1 \cdot 3 \cdot 5 \cdot 8 = 1 \cdot 4 \cdot 5 \cdot 6 = 2 \cdot 3 \cdot 4 \cdot 5$. After his question is answered the visitor states that he can deduce the exact number of children in each family. Therefore the answer must have been "no" and the numbers we seek are 2, 3, 4, and 5.

Note:

This problem was originated by L. R. Ford, formerly head of the Department of Mathematics, Illinois Institute of Technology, Chicago.

96. Red's two statements: "I did not kill Don" and "Shorty lied when he said I'm guilty," are either both true or both false. Since it is given that only one statement is false, they are both true. Therefore one of Red's other two statements is false.

Shorty's false statement was: "Red is the guilty man." Therefore his other three statements were true.

Joey's statement: "I never saw Shorty before," contradicts one of Shorty's true statements and is therefore false. His other three statements were therefore true, including the fact that he did not kill Donavan and the fact that Red has never been in Trenton.

Since Red told the truth about not having been in Trenton, his false statement must have been that he never saw Hank before.

Tony's last three statements have been shown to be true. Therefore it was false that Hank lied about owning a revolver.

Hank told the truth about not owning a revolver, about knowing Red, and about being in Philadelphia. Therefore he lied when he said: "I did not kill Donavan." Thus Hank is guilty.

97. The motorist knows that the man standing at the fork of the road *always* tells the truth or *always* lies depending on whether he is a Euphorian or a non-Euphorian. Knowing this the motorist is allowed to ask one question for the purpose of determining which of the two roads leads to Avon. The question can take many forms. One form could be, "If I were to ask you 'Does the right-hand road lead to Avon,' would you say yes?" This question will elicit the correct answer whether the respondent is a Euphorian or not.

For a similar problem worked out by symbolic logic, refer to [24], p. 16, Problem 2.

98. We know that Sir Barnacle owns the *Gabrielle;* Mr. Collingwood, the *Lorna;* Mr. Nelson, the *Rosalind;* and Colonel Drake, the *Melissa;* so Dr. Benbow must own the *Thalia.* Dr. Benbow cannot be Gabrielle's father because he would own a yacht named after his own daughter and each man has named his yacht after a daughter of one of the others. We know that Thalia is Mr. Collingwood's daughter and Melissa is Sir Barnacle's daughter. Therefore Gabrielle must be the daughter of either Colonel Drake or Mr. Nelson. If we assume that Gabrielle is Colonel Drake's daughter, then Lorna must be Mr. Nelson's daughter since Thalia, Gabrielle, and Melissa are accounted for and the name of Mr. Nelson's yacht eliminates Rosalind. This identifies Rosalind as Dr. Benbow's daughter and leads to the conclusion that Gabrielle's father

(Colonel Drake) owns the *Rosalind.* This conclusion contradicts the fact that Colonel Drake owns the *Melissa,* and so Gabrielle cannot be Colonel Drake's daughter, and must instead be the daughter of Mr. Nelson. Since Gabrielle's father owns the yacht named after Dr. Benbow's daughter, we see that Rosalind's father is Dr. Benbow. We conclude, then, that Lorna's father is Colonel Drake.

99. From statement (3), the *Washington,* having sails, is the sloop. From (2), the *Lincoln* is not the steamship. Therefore the *Lincoln* is the oil tanker, and the *Jefferson* is the steamer.
Consider the following table:

	Berm.		**Bost.**		**Hal.**		**Lond.**		**N. N.**		**N. Y.**	
	L.	**(A.)**	**L.**	**A.**	**L.**	**A.**	**L.**	**A.**	**L.**	**A.**	**L.**	**A.**
Washington-sloop	x											
Lincoln-oil tanker	x	x										
Jefferson-steamer	x											

L means "leaves from," and **A** means "arrives at."

From (1), the oil tanker is not going to Bermuda. Therefore place an **x** at the intersection of *oil tanker* and **A**. Since Bermuda is an arrival port, circle the **A** in the second Bermuda column, and place crosses under all **L**'s of Bermuda. These are shown above.

From (2), the *Lincoln* arrives at Halifax. Therefore circle the proper box and place crosses in all other **L** and **A** boxes of Boston, London, Newport News, and New York. Also from (2) circle: steamship-left London; and place **x**'s in the boxes thereby eliminated.

There should now be no trouble completing the table using the clues in (3). Note that one ship did go to Boston, and that it had to be the *Jefferson.*

In summary: the *Washington* was the sloop, and went from New York to Bermuda. The *Lincoln* was the oil tanker and went from Newport News to Halifax. The *Jefferson* was the steamer and went from London to Boston.

100. From (2), John is a senior in medicine. From (3), neither Karl nor Luke is in engineering. From (4), Oswald is the graduate student in law. From (5), Karl is not a sophomore and not in pharmacy.

This leaves only dentistry for Karl and pharmacy for Luke, who is therefore the sophomore. Finally, from (1), Karl is the freshman, and Mark is the junior in engineering.

101. From (1), Todd won as a heavyweight. From (2), Grant and Slater were winners and Mitchell a loser. From (3), O'Brien was a loser in the 145-lb. class, and Ford lost in the 118-lb. class. From (1), Ford's loss was to Brown.

If these have been recorded in tabular form it will be seen that four winners and three losers have been accounted for. Hence Robinson, not yet accounted for, must be a loser.

From (2), Grant defeated Mitchell who, by elimination, was in the 126-lb. class. This leaves Slater as winner in the 145-lb. class, and Robinson, the heavyweight loser.

In summary:

118-lb.: Brown beat Ford;
126-lb.: Grant beat Mitchell;
145-lb.: Slater beat O'Brien;
Heavyweight: Todd beat Robinson.

102. In the following table, the headings at the top indicate that the guilt of each is considered in turn. The headings at the left indicate that the truth (**T**) or falseness (**F**) of each one's statement is considered on the assumption of the guilt of each in turn.

	J	H	T	E	D
JOHN	**F**	**T**	**T**	**F**	**F**
HENRY	**T**	**F**	**T**	**F**	**T**
THOMAS	**F**	**F**	**F**	**T**	**F**
DAVID	**T**	**T**	**F**	**F**	**T**
ERNEST	**F**	**F**	**T**	**T**	**F**

For example, if John were guilty, John's statement would be false; Henry's would be true; Thomas's would be false; David's would be true; Ernest's would be false. These are all indicated in the first column. The others were obtained in the same way. The only column which indicates that just three sons were telling the truth is the third. Therefore Thomas was guilty.

Note:

(i) After the top two rows are filled in, the third row can read **T** only if two **F**'s occur above it. Therefore Thomas would be telling the truth only if Ernest were guilty. The fourth row can read **T** only if one **F** and one **T** occur in the top two rows. The fifth row must read opposite of the fourth row.

(ii) These results can be interpreted in terms of Truth Tables in the subject of symbolic logic. References: [24], and Allendoerfer, C. B., and Oakley, C. O. *Principles of Mathematics*. New York: McGraw-Hill Book Co., Inc., 1955. Chapter I.

103. From (3), Joe is not 8 or 9; the 9-year-old is not in one of the two top stories; the 8-year-old is not in one of the two bottom stories; Joe is not on the very top nor the very bottom floor.

From (4), Jack is not the 5-year-old; Jack is not on the bottom floor; the 5-year-old is not on the top floor.

From (5), Jack is not the youngest; Tom is not the oldest, nor the second oldest. Also Tom can be 5 if Jack is 10, 9, or 8; but he can only be 8 if Jack is 10.

All these eliminations leave very few possibilities to try, and a tabular arrangement soon suggests the solution: Jack is 8; Joe is 10; John is 9; and Tom is 5. They live on the fourth, third, second, and first floors, respectively.

104. Subtracting the assigned totals in conditions 1 to 3 from the entire total of time and number, viz. 56, we obtain 12 as the combined total for two runners with unassigned totals. Dividing 12 into totals 9 and 3, or 10 and 2, is ruled out by condition 2. Totals 8 and 4 are not possible because 4 has already been assigned. Thus 6 and 6, or 7 and 5, remain; and the complete set of totals is 13, 9, 9, 9, 7, 5, 4; or 13, 9, 9, 9, 6, 6, 4.

To obtain total 13, either number 7 comes in 6th or number 6 comes in 7th. In the latter case number 7 can combine only to total 9—that is, he must come in second. But condition 4 rules this out. Therefore, number 7 comes in 6th, and only number 2 can come in seventh.

Second place cannot go to number 6 ($2+6=8$), or to number 5 (6 would then have to come in first, by condition 5, giving two totals of seven). Therefore 3 or 4 will be second. By similar reasoning 5 or 6 will be first. Since places 3, 4, 5 remain and totals 4, 9, 9 still remain, we must assign third place to 1, fourth place to 5, and fifth place to 4. The rest follows readily giving the final finish as: 6, 3, 1, 5, 4, 7, 2.

105. Indicate the missing digits by the letters: $a, b, \ldots k$

$$\begin{array}{r} 4\,a\,b \\ 3\,c \\ \hline 3\,6\,d\,e \\ f\,g\,7\,h \\ \hline i\,j\,3\,k\,e \end{array}$$

(i) $499 \cdot 39 = 19{,}461$ is the largest product obtainable. Therefore $f = i = 1$.

(ii) Since $6+7$ sums to 13, $d+h \leqq 9 = k$.

(iii) From (i), there was nothing to carry in $3+g+1=j$. Hence $g \leqq 5$.

(iv) From (iii), $3(400+10a+b) \leqq (1570+h)$.

(v) Since $499 \cdot 7 = 3493 < 36de$, $c=8$ or 9.

(vi) If $c=9$, $9(400+10a+b) = 3600+90a+9b \leqq 3699$. Therefore $10a+b \leqq 11$. But this is impossible because $3(10a+b)$ ends in $70+h$. Hence $c=8$.

(vii) Therefore $8(400+10a+b) < 3699$, and $10a+b \leqq 62$.

(viii) A few trials now yield the answer $457 \cdot 38 = 17{,}366$.

106. Number the columns of the sum as follows:

$$\begin{array}{ccccc} \underline{1} & \underline{2} & \underline{3} & \underline{4} & \underline{5} \\ A & H & A & H & A \\ & T & E & H & E \\ \hline T & E & H & A & W \end{array}$$

From (3) and (5): $H = W+1$, and from (4), $H+H > 10$.
Therefore: $H = 6, 7, 8$, or 9.
Whence, from (4): $A = 2, 4, 6$, or 8.
Then, from (3): $E = 3, 2, 1$, or 0.
From (2): $T = 7, 5, 3$, or 1.
From (1): $A+1 = T$, which holds only when $H=7$, $A=4$, $E=2$, and $T=5$.
The given addition is: $47{,}474 + 5272 = 52{,}746$.

107. **(i)** Since $T-S=T$, $S=0$.

(ii) Since $V-L=L$, $2L=V$, whence for $L=1, 2, 3, 4$, $V=2, 4, 6, 8$.

(iii) $R \times GR$ $(=L\,T\,R)$ ends in R. Therefore $R=1$ or 5 or 6. But $R \neq 1$, because $1 \times G1$ equals a three-digit number, a contradiction.

(iv) Assume $R=6$. From $I \times GR = V\ L\ S$, which ends in 0, I must be 5. $G6 \times 5 = 210$, or 420, or 840, because $V = 2\ L$; (not 630, for then R and V would both be 6). Therefore $(10G+6) \times 5 = 210$, 420, 840, from which $50\ G = 180$, 390, 810, all of which are impossible for integral G. Hence $R=5$, and I is an even number ($\neq 0$).

(v) By a similar method we test all possibilities of the pattern $G5 \times I = 210$, 420, 630, 840, for $I = 2$, 4, 6, 8, and find that the only possibility is $35 \times 6 = 210$. Therefore $G=3$, $R=5$, $I=6$, $V=2$, $L=1$, $S=0$.

(vi) The rest follows readily, and we soon discover that the divisor is 35, the dividend 19,775, and the quotient 565.

108. By the distributive property, 1999 times $abcd = 2000abcd - abcd$. This can be found by doubling $abcd$, placing 3 zeros at the end, and then subtracting $abcd$. If, in this process, 4 repeated digits are found at the end of the difference, $abcd$ can be determined by working backwards. Here is the reconstruction for repeated digit 1.

(1)
$$\begin{array}{rr} & \ldots\text{–}000 \\ - & \text{– – – –} \\ \hline & \ldots 1111 \end{array}$$

(2)
$$\begin{array}{rr} & \ldots\text{–}000 \\ - & \text{–}889 \\ \hline & \ldots 1111 \end{array}$$

(3)
$$\begin{array}{rr} & \ldots 8000 \\ - & \text{–}889 \\ \hline & \ldots 1111 \end{array}$$

(4)
$$\begin{array}{rr} & \ldots 8000 \\ - & 6889\ (=abcd) \\ \hline & \ldots 1111 \end{array}$$

By this method we obtain for repeated digits 1, 2, 3, 4, 5, 6, 7, 8, 9, $abcd$ is 6889; 3778; 0667; 7556; 4445; 1334; 8223; 5112; or 2001. Since $a \neq 0$, delete 0667.

109. Following the pattern in (108):

$$\begin{array}{rcccc} \ldots & 2d & 0 & 0 & 0 \\ - & a & b & c & d \\ \hline \ldots & t & t & t & t \end{array}$$

Hence $t+d=10$, $t+c=t+b=9$, and $t+a+1=2d$, or $2d-10$, or $2d+10$. Therefore $b=9-t$, $c=9-t$, $d=10-t$; and $a=9-3t$ for $t=1$, 2, 3; and $a=19-3t$ for $t=4$, 5, 6; $a=29-3t$ for $t=7$, 8, 9.

Notes:

(i) The problem might have been worded so that the product ended with six repeated digits. $abcd$ would then equal 1334 or 2001, for endings of six 6's or six 9's.

(ii) John Middleditch of Tucson, Arizona, devised the formula:

$$a = 10\left[\frac{t+2}{3}\right] - 3t - 1,$$

where "[ ]" stands for the "greatest integer" function.

(iii) The solution can also be given as $a \equiv 9 - 3t$ (mod. 10). Reference: [16], Chapter 4.

V

MISCELLANEOUS PROBLEMS

110. The various combinations of membership on teams can be exhibited conveniently in a Venn diagram.

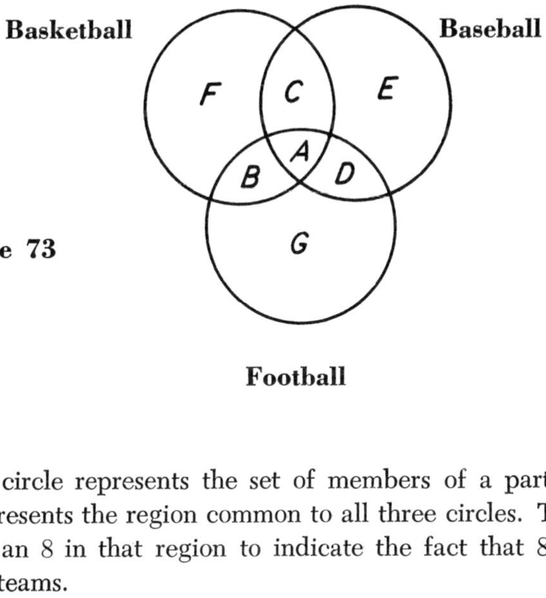

Figure 73

Each circle represents the set of members of a particular team. A represents the region common to all three circles. Therefore we write an 8 in that region to indicate the fact that 8 play on all three teams.

Since the union of A and B represents those who play both football and basketball, write a 4 in region B so that A and B together will make 12, as given. Continue in this manner, considering the data in reverse of the order given in the statement of the problem. The final diagram will look like figure 74. The sum of the numbers in the various regions is 43, which is therefore the total number of players.

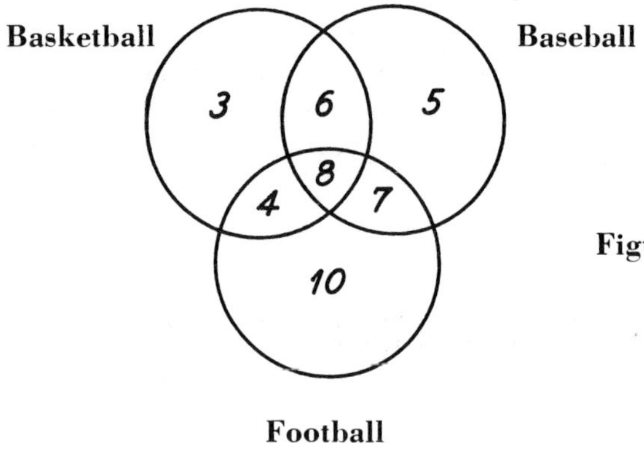

Figure 74

111. If the integer sought is represented by x, then $\frac{x-2}{3}=p$, $\frac{x-3}{5}=q$, $\frac{x-2}{7}=r$. We try $x=9$, which is the smallest positive value of x that makes r integral. Since this does not work for p and q, we try higher values, using the principle that if $ax+b$ is divisible by c, then increasing (or decreasing) x by a multiple of c yields a result that is also divisible by c.

If we increase 9 by consecutive multiples of 7, we find 23 $(=9+2\cdot 7)$ makes p integral. It happens also to make q integral, and is therefore one number that we seek. If we now increase 23 by multiples of 21 $(=3\cdot 7)$, p and r will remain integral. We find $x=128$ $(=23+5\cdot 21)$ makes q integral. The required numbers are therefore 23 and 128.

Note:

(i) This problem was first stated by Sun-Tsu in the first century A.D. A generalization of the problem, based on a rule stated by Sun-Tsu, is known in the *theory of numbers* as the Chinese Remainder Theorem. Reference: [16], page 79.

(ii) As a special case of the Chinese Remainder Theorem, let us find the smallest integer which gives a remainder of 1 when divided by each of 4, 5, 6. To obtain the answer, find the lowest common multiple of 4, 5, 6, viz. 60. The required number is therefore 61. All numbers that meet these conditions are represented by $60n+1$.

As a second example, find the smallest integer which gives a remainder of 1, 2, 4, 6, 10 when divided by 2, 3, 5, 7, 11 respectively. We note that in each case the remainder is 1 less than the corresponding divisor. Hence the smallest integer that meets the given conditions is $2 \cdot 3 \cdot 5 \cdot 7 \cdot 11 - 1$, or 2309. All such numbers are represented by $2310\,n-1$.

112. $2^n \cdot 3^{2n} - 1 = 2^n \cdot 3^n \cdot 3^n - 1 = 6^n \cdot 3^n - 1 = 18^n - 1$. Now $a^n - 1 = (a-1)(a^{n-1} + \ldots + 1)$; hence $18^n - 1 = (18-1) \times (\text{integer})$. Therefore the given expression is divisible by 17.

Note:

For other problems of this type see [18], pp. 356-57, and [4], chapter XXXV.

113. Let p = the number of pennies, n = the number of nickels, and d = the number of dimes. Then $p + 5n + 10d = 162$ and $n = \frac{p+n+d}{3}$. If we eliminate n from these two equations, we obtain:

$$p = \frac{54 - 5d}{2}.$$

Since 54 is an even number, d must be even for p to be integral. We can now make the following table:

d	0	2	4	6	8	10
p	27	22	17	12	7	2
n	27	24	21	18	15	12

But since the statement of the problem implies that there must be at least one of each type of coin, we omit the first column.

114. The kth group ends in $1 + 2 + \ldots + k = \frac{k(k+1)}{2}$. Since there are k numbers in this group, its first term is $\frac{k(k+1)}{2} - k + 1$. Substituting in the formula $S = \frac{n(a+l)}{2}$, we obtain:

$$S = \frac{k}{2}\left[\frac{k(k+1)}{2} - k + 1 + \frac{k(k+1)}{2}\right] = \frac{k(k^2+1)}{2}.$$

Note:

This sum is the same as the constant for a magic square, k by k. See the solution of Problem 28.

115. Let the five-by-five square be checkered as in Fig. 75.

Figure 75

In accord with the teacher's instructions, if a student is in a seat that corresponds to a black square, he must go to one that corresponds to a white square. But this is impossible, because there are 13 black squares and only 12 white squares. Hence the teacher's order cannot be carried out.

Note:

The technique of using checkered squares may be applied to other problems. The following is an example:

Suppose two diagonally opposite corner squares of a regulation checkerboard are removed. Is it possible to cover this board (as in a jig-saw puzzle) with 31 rectangles, each 2 squares by 1 square?

To solve this, look at the colors of two squares that might be removed. You will find that they are both of the same color. Now consider the fact that when a 2-by-1 rectangle is placed on the board, it covers one square of each color. You should therefore find that the proposal cannot be carried out.

Reference: Golomb, S. W. *Polyominoes.* New York: Charles Scribner's Sons, 1965.

116. (1) In each of the following, refer to the letters in the diagrams and move the piece indicated according to the rules. Underlined letters are black moves.

Game 1: $\underline{a}$, c, $\underline{b}$; 3 moves.
Game 2: $\underline{b}$, d, e, $\underline{c}$, $\underline{a}$, b, d, $\underline{c}$; 8 moves.
Game 3: $\underline{c}$, e, f, $\underline{d}$, $\underline{b}$, $\underline{a}$, c, e, g, $\underline{f}$, $\underline{d}$, $\underline{b}$, c, e, $\underline{d}$; 15 moves.

(2) The general plan can be seen by referring to the number of consecutive moves each side makes at a time in (1) above:

(B = Black; W = White)
Game 1: B1, W1, B1
Game 2: B1, W2, B2, W2, B1
Game 3: B1, W2, B3, W3, B3, W2, B1

For n pieces on each side, the number of moves is $S=1+2+3+\ldots+(n-2)+(n-1)+n+n+n+(n-1)+(n-2)+\ldots+3+2+1$. This consists of two sums, each $1+2+\ldots+n$, added to another term n which occurs in the middle. Hence $S=\frac{2n(n+1)}{2}+n=n^2+2n$. Note that for $n=1, 2, 3$, $S=3, 8, 15$ as in the above solutions.

(3) Consider Game 4:

Figure 76

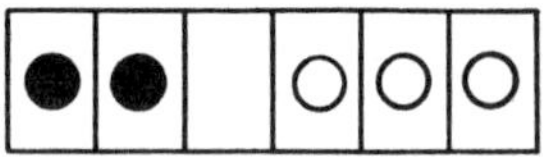

To interchange places, each B must make 4 single moves to the right. Since there are 2 B's, they must make $2\times4=8$ single moves. Similarly, each of the W's 3 pieces must make 3 single moves to the left, or $3\times3=9$ single moves. But some of these moves are saved by the fact that every jump (either way) promotes the solution by 2 moves. Therefore, the theoretical minimum $M=8+9-$number of jumps. Each of the 2 B's must encounter each of the 3 W's, and regardless of which jumps which, there must be $2\times3=6$ jumps. Therefore $M=8+9-6=11$ moves.

In general consider the game with b blacks and w whites:

Figure 77

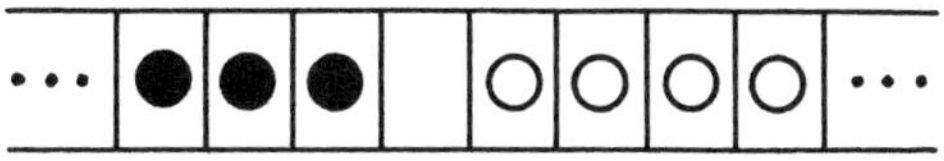

To interchange places: each B must make $w+1$ single moves $=b(w+1)$, each W must make $b+1$ single moves $=w(b+1)$. From their sum subtract the number of jumps $=bw$. Thus the theoretical minimum is $b(w+1)+w(b+1)-bw=b+w+bw$, which is the required formula.

Note that if $b=w=n$, we obtain $n+n+n^2$ or n^2+2n, confirming the formula in (2) above.

117. **(a)** (*i*) When $N=1$, A removes 1 and wins.

(*ii*) When $N=2$, A (must) remove 1, leaving $N=1$, and B the first move. Therefore B wins as in (*i*).

(*iii*) When $N=3$, A (must) remove 1, leaving $N=2$. Thus A —now the second player—wins as in (*ii*).

(*iv*) When $N=4$, A removes 2 matches and wins as in (*ii*).

(*v*) When $N=5$, A removes 1 or 2 matches, leaving $N=4$, 3. Thus B wins as in (*iv*), (*iii*).

(*vi*) When $N=6$, A removes 1 match, leaving $N=5$, and wins as in (*v*).

(b) *If* $N_1, N_2, \ldots N_x$ are the first x winning positions, it is clear from the pattern in the special cases in (**a**), that

$$N_{x+1}=2N_x+1. \text{ Also } N_x=2N_{x-1}+1.$$

$$\text{Hence } 2N_x+1=2^2N_{x-1}+2+1. \text{ Also } N_{x-1}=2N_{x-2}+1.$$

$$\text{Hence } 2^2N_{x-1}+2+1=2^3N_{x-2}+2^2+2+1.$$

Continuing in like manner, we obtain finally:

$$N_{x+1}=2^xN_1+2^{x-1}+2^{x-2}+\ldots+2+1.$$

But $N_1=2$ [case (*ii*) of part (**a**)] and $2^{x-1}+2^{x-2}+\ldots+2+1=\dfrac{2^x-1}{2-1}=2^x-1$. Therefore $N_{x+1}=2^{x+1}+2^x-1=2^x(2+1)-1=3\cdot2^x-1$. Thus for $x=$ 0, 1, 2, 3, 4, we obtain $N_{x+1}=2, 5, 11, 23, 47$.

Note:

For the rules and theory of the original version of the game of NIM and some of its interesting variations, see [1], [13], and [27].

118. **(a)** If the hundreds, tens, and units digits of a three-digit number are a, b, c, the sum of the six permutations is: $(100a+10b+c)+(100a+10c+b)+(100b+10a+c)+(100b+10c+a)+(100c+10a+b)+(100c+10b+a)$. This sum is $222a+222b+222c$, whose average is $\dfrac{222(a+b+c)}{6}=37(a+b+c)$. Since the latter is an integer, the first part is proved.

(b) $37\,(a+b+c) = 36\,(a+b+c)+a+b+c =$ (multiple of 9)+S, where S is the sum of the digits. If S is a one-digit number, then $\frac{37(a+b+c)}{9}$ gives a remainder of $a+b+c$. If S is a two-digit number, say $10m+n$, then $37(a+b+c)=36(a+b+c)+9m+m+n$. In this case, $\frac{37(a+b+c)}{9}$ gives a remainder of $m+n$, which is the continued sum of $a+b+c$. Obviously this can be continued as long as necessary until a one-digit sum is obtained. Therefore the second observation must be modified to say: Take the continued sum of the digits until a one-digit sum is obtained; then this sum will be equal to the sum of the digits of the original three digits.

Notes:

(i) That the statement in (**b**) is not true can be shown by a special case. For example, if $a=1$, $b=2$, $c=4$, then $a+b+c=7$; but $37(a+b+c)=259$ with $2+5+9=16$. The proof in (**b**) shows that we must find the continued sum: $1+6=7$.

(ii) An examination of the proof of this problem will show that the result of the continued addition is the same as that obtained as the least positive (or zero) remainder, after dividing by 9. For example: $\frac{37462}{9}$ leaves a remainder of 4; and $3+7+4+6+2=22$, $2+2=4$.

This fact has a number of applications: (1) as a test for divisibility by 9 (see solution to Problem 124); (2) as a device for checking simple arithmetic calculations. In the second application it is known as *Casting Out Nines* and has been in use for several hundred years as a checking device in business accounting. For example: the sum of 58,372 and 26,419 is 84,791. If we *cast out nines* for each number, 58,372 becomes 7, 26,419 becomes 4, and 84,791 becomes 2. The sum of 7 and 4, or 11, also becomes 2. This device is necessary, but not sufficient for the accuracy of the result. It will work also for subtraction, multiplication, and with modification, for division.

(iii) It is possible to cast out 99's, or 999's, etc. For a mathematical recreation based on this see:

CHAROSH, MANNIS. "On Casting Out 999 . . .'s and Certain Cyclic Permutations." *Scripta Mathematica:* March 1941. p. 47.

119. The suggested generalization is: $x^3 = S^2 - (S-x)^2$, where x is any positive integer, and S is the sum of the integers from 1 to x.

Since S is the sum of an arithmetic progression of x terms with first term 1 and difference 1, $S = \frac{x+x^2}{2}$. We must therefore prove:

$$x^3 = \left(\frac{x+x^2}{2}\right)^2 - \left(\frac{x+x^2}{2} - x\right)^2.$$

The right side of this reduces to x^3, as required. The simplest way of evaluating it is to factor it as the difference of two squares.

Note:

A formula to be found in many books on advanced algebra is the following: $1^3+2^3+3^3+ \ldots +n^3 = (1+2+3+ \ldots +n)^2$. (See for example [16], page 50.) The proof of the required result can then be given in one step by observing that:

$$n^3 = (1^3+2^3+ \ldots +n^3) - [1^3+2^3+ \ldots +(n-1)^3].$$

120. Let the consecutive ages of the man be $x, x+1, \ldots, x+5$, and the corresponding ages of his grandson: $y, y+1, \ldots, y+5$. Since $x+a$ is a multiple of $y+a$ for $a=1, 2, \ldots, 5$, then $(x+a)-(y+a)$ or $x-y$ is also a multiple of $y+a$.

If $y=2$, $x-2$ is the lowest common multiple of 2, 3, 4, 5, 6, and 7, which is 420. Therefore $x-2=420$, and $x=422$, an impossible age.

But if $y=1$, $x-1$ is the lowest common multiple of 1, 2, 3, 4, 5, 6, or 60. Therefore, $x=61$, and man and grandson are respectively 66 and 6.

121. Since the light goes on at 12:00 and is off at 12:09, the interval must be less than 9 minutes. If the interval is 2 or 4 minutes, the light would be on at 12:09; if 1, 3, or 5 minutes, the light would be off at 12:17; if 6 or 8 minutes, it would be off at 12:58. Thus the interval can only be 7 minutes long. This checks, and the light is off at 2:00.

Note:

There are many sets of three or more times which will determine uniquely the interval of change. It may interest readers to seek other such sets and perhaps to determine a procedure for finding them.

122. A formula will be suggested by taking special cases and observing the results. For example in a rectangle 6 by 4, $I=15$, $S=20$, $A=24$; in a rectangle 6 by 10, $I=45$, $S=32$, $A=60$; and in a rectangle 5 by 8, $I=28$, $S=26$, $A=40$. A study of these results suggests that $A=\frac{S}{2}+I-1$. To verify this formula, note first that S will equal the perimeter P, since there is a lattice point at the beginning of a segment. If L and W represent the length and width of the rectangle, then $P=2(L+W)$ and $P=S$. Therefore, $\frac{S}{2}=L+W$. Since there are $L+1$ lattice points on the length (L beginning points of unit segments plus one end point), then there are $L-1$ lattice points on the length, not counting the beginning or end. Likewise the number of lattice points on the width, not counting the beginning or end, is $W-1$. The intersection of these two is $(L-1)(W-1)$, which is therefore the number of interior lattice points of the rectangle. Substituting in the suggested formula:

$$A=\frac{S}{2}+I-1$$

$$A=L+W+(L-1)(W-1)-1$$

$$A=L+W+LW-W-L+1-1$$

$A=LW$. This verifies the suggested formula. The argument is reversible.

Note:

The generalization of this theorem in a plane is known as Pick's Theorem. It states that *the area of a polygon whose vertices are lattice points is equal in magnitude to one-half the number of lattice points on its boundary plus the number of interior lattice points minus one.* For a proof see [45].

The reader might also try to generalize this theorem for three dimensions.

123. (a) **First Method**

If x is the divisor and PQR (representing $100P+10Q+R$) is the quotient, then $Px=1296$, $Qx=432$, and $Rx=864$. Therefore $P=3Q$ and $R=2Q$. Also $Q=$ 1, or 2, or 3, since $P\leq 9$. Thus:

if $Q=1, P=3, R=2, PQR=312, x=432$;

if $Q=2, P=6, R=4, PQR=624, x=216$;

if $Q=3, P=9, R=6, PQR=936, x=144$.

The dividend $=312\times 432=624\times 216=936\times 144=134{,}784$.

Second Method

The dividend may be determined by working backwards:

$$
\begin{array}{r@{\;}l}
 & \quad P\,Q\,R \\
\cline{2-2}
x\;) & \quad y\uparrow\uparrow \\
 & 1296 \\
\cline{2-2}
 & \quad\; u \\
 & \quad 432 \\
\cline{2-2}
 & \qquad v \\
 & \quad\; 864 \\
\cline{2-2}
\end{array}
$$

Since the division is exact, $v=864$, and 4 is the last digit of dividend y. We now have:

$$
\begin{array}{r}
u \\
432 \\
\hline
864 \\
864 \\
\hline
\end{array}
$$

Since $u-432=86$ (the last digit 4 is carried down *after* the multiplication), $u=518$. Continuing in this way, we determine finally that $y=134{,}784$. By considering the factors common to 864, 432, and 1296, the possible values of x, P, Q, and R may be determined.

(b) Using the method of the second solution in (**a**), we find the dividend to be $100A+10B+C$.

Note:

Consider the following division:

$$
\begin{array}{r@{\;}l}
 & \qquad 3012 \\
\cline{2-2}
432\;) & 1301184 \\
 & 1296 \\
\cline{2-2}
 & \quad\; 518 \\
 & \quad\; 432 \\
\cline{2-2}
 & \qquad 864 \\
 & \qquad 864 \\
\cline{2-2}
\end{array}
$$

The method of working this backwards is the same as in the second solution in (**a**) except that the 518 was placed with the 5 under the 6, instead of under the 9. By varying the indentions, a number of new divisions are determined. In each, the given conditions are met except for the zeroes (e.g. $0\times 432=0$), which technically should be subtracted, but do not appear in the algorithm. Thus if zeroes are permitted, the quotients in (**a**) are: $AOBC$, $AOOBC$, $ABOC$, and $ABOOC$, in each of which $(A, B, C)=(4, 3, 2)$, or $(2, 1, 6)$, or $(1, 4, 4)$; and "0" is zero.

In (**b**), if zeroes are permitted, the new dividends are: $1000A+10B+C$, $10{,}000A+10B+C$, $1000A+100B+C$, and $10{,}000A+1000B+C$.

124. For convenience, we will use the symbol: $M(n)$ to represent *some multiple of n.* We now state the generalization as follows: If $10R+U=M(7)$, then $R-2U=M(7)$, and conversely. R may have any number of digits.

Proof: Since $10R+U$ and $21U$ are each $=M(7)$, $10R+U-21U$, or $10(R-2U)=M(7)$. And since 10 is not a multiple of 7, $R-2b=M(7)$. Since the argument is reversible, the converse is also true.

Notes:

(i) The intention of the problem is to suggest a test for divisibility by 7. Thus to test 53,844, take the following steps: $5384-(2\cdot 4)=5376$; $537-(2\cdot 6)=525$; $52-(2\cdot 5)=42$. Since $42=M(7)$, then $53{,}844=M(7)$.

(ii) Tests for divisibility by other numbers may be devised by using the following principles:

(*a*) If $M(n)$ is added to, or subtracted from $M(n)$, the result is also $M(n)$.

(*b*) The product of $M(n)$ by any integer is also $M(n)$.

(*c*) If $AB=M(p)$ (p is a prime) and $A\neq M(p)$, then $B=M(p)$.

(*a*) and (*c*) were used in the proof above.

(iii) Here are some other tests for divisibility: Any integer may be represented as follows: $N=a_0+10a_1+100a_2+\ldots.$

Test for divisibility by 2^n

Since $N=a_0+2(5a_1+50a_2+\ldots)$, $N=M(2)$ if $a_0=M(2)$.
Since $N=a_0+10a_1+4(25a_2+\ldots)$, $N=M(4)$ if $a_0+10a_1=M(4)$.
Since $N=a_0+10a_1+100a_2+8(\ldots)$, $N=M(8)$ if $a_0+10a_1+100a_2=M(8)$.
In general, N will be a multiple of 2^n if the number consisting of the last n digits of N is divisible by 2^n. Thus $375{,}299{,}685{,}832=M(8)$, because $832=M(8)$.

Test for divisibility by 9 or by 3

Since $N = a_0 + a_1 + a_2 + \ldots + 9(\ldots)$, $N = M(9)$ or $M(3)$ if the sum of the digits of $N = M(9)$ or $M(3)$, respectively.

Following the idea in the proof of the original problem, we can devise other tests.

Test for divisibility by 11

If $10R + U = M(11)$, then $10R + U - 11R$, or $-R + U = M(11)$. Therefore $R - U = M(11)$. Thus to test 534182, we subtract as follows: $53418 - 2 = 53416$; $5341 - 6 = 5335$; $533 - 5 = 528$; $52 - 8 = 44 = M(11), \ldots$.

This is equivalent to the following rule: If the sum of the even-placed digits of N is E, and the sum of the odd-placed digits is F, then $N = M(11)$ if $E - F = M(11)$.

Test for divisibility by 17

If $10R + U = M(17)$, then in turn, $10R - 16U$, $5R - 8U$, $5R - 8U - 17U = 5R - 25U$, $R - 5U$ are each $= M(17)$. To test 1428, subtract as follows: $142 - 40 = 102$; $10 - 10 = 0 = M(17)$, etc.

The reader may try to prove the following divisibility tests. Since the coefficient of R in each was made equal to *one*, the tests are convenient to apply.

$$
\begin{aligned}
N = 10R + U &= M(13), \text{ if } R + 4U = M(13);\\
&= M(19), \text{ if } R + 2U = M(19);\\
&= M(23), \text{ if } R + 7U = M(23);\\
&= M(31), \text{ if } R - 3U = M(31);\\
&= M(101), \text{ if } R - 10U = M(101).
\end{aligned}
$$

In general, $10R + U = M(p)$, if $bR - aU = M(p)$ and $10a + b = M(p)$ (b and p have no common factor).

Proof: Each of the following in turn $= M(p)$: $bR - aU$, $10bR - 10aU$, $10bR + bU - 10aU - bU$, $b(10R + U) - U(10a + b)$, $b(10R + U)$, and finally, since b and p have no common factor, $10R + U = M(p)$.

(iv) It is possible to reduce the work in a divisibility test further. For example: $100A + B = M(7)$, if $2A + B$, or $2A + 8B$ or $A + 4B$, or $A - 3B = M(7)$. Thus to test 58,261, subtract as follows: $582 - 3 \cdot 61 = 399 = M(7)$; therefore 58,261 is a multiple of 7.

(v) Divisibility tests may thus be devised that work with any desired number of the last digits of N. The greatest generalization we give here is the following:

$$\text{If } N = \underbrace{A}_{a\text{ digits};}\ \Big|\ \underbrace{B}_{b\text{ digits}} = 10^bA+B, \text{ then } N=M(p) \text{ if:}$$

(a) $rA - sB = M(p)$;

(b) $10^bs + r = M(p)$;

(c) r and p have no common factor.

The proof is similar to that of the previous generalization. Each in turn is $=M(p)$: $rA - sB$, $10^brA - 10^bsB$, $10^brA + rB - 10^bsB - rB$, $r(10^bA + B) - B(10^bs + r)$, $r(10^bA + B)$, and finally, since r and p have no common factor, $10^bA + B$, or N.

As a final application: prove that if $N = 1000A + B = M(13)$, then $A - B = M(13)$.

125. In general, for a square n by n, the sums in the two diagonals are equal. The numbers in each form an arithmetic progression.

Beginning at the upper left, the first term is 1, the last term is n^2, and the number of terms is n. Hence the sum is $\dfrac{n(n^2+1)}{2}$

Beginning at the upper right, the first term is n, the last term is $n^2 - (n-1)$, and the number of terms is n. Hence the sum is $\dfrac{n(n+n^2-n+1)}{2} = \dfrac{n(n^2+1)}{2}$

Figure 78

1	15	14	4
12	6	7	9
8	10	11	5
13	3	2	16

Thus the two sums are equal. This sum is the one required for a magic square n by n (see the solution to Problem 28).

The magic square can be completed by interchanging each pair of numbers that are in cells symmetric with respect to the center of the square, except those in the diagonals (see Fig. 78). The magic sum, 34, will be found not only vertically, horizontally, and diagonally, but also in a great number of symmetric patterns. Here are a few: $1+4+16+13$; $1+15+6+12$; $6+7+11+10$; $1+9+8+16$; $1+3+14+16$; $15+7+10+2$; and many others which the reader may wish to find.

For more information about magic squares, see [1], [13], [27], and the following:

ANDREWS, W. S. *Magic Squares and Cubes.* Chicago: Open Court, 1917.

USPENSKY, J. V., and HEASLET, M. A. *Elementary Number Theory.* New York: McGraw-Hill Book Co., Inc., 1939.

SITOMER, HARRY. "The Magic of Magic Squares." *Mathematics Student Journal:* January 1964.

126. If .5 is subtracted from each term of the series, we obtain 2.5, -2.5, $+2.5$, -2.5, . . . , which is a geometric progression with a constant ratio of -1. The nth term of this series is $2.5(-1)^{n-1}$. Therefore the nth term of the given series is obtained by restoring the .5 previously subtracted. Thus a correct formula is: $2.5(-1)^{n-1}+.5$.

Of the many formulas submitted by solvers, some of the more interesting were:

(1) $\left(-\frac{5}{2}\right)\cos n\pi+\frac{1}{2}$;

(2) $3-5\left(\cos\frac{n\pi}{2}\right)^2$;

(3) $5n-2-10\left[\frac{n}{2}\right]$, where $\left[\frac{n}{2}\right]$ is the greatest integer contained in $\frac{n}{2}$.

(4) For the general series: $a+b+a+b+\ldots$ in which the a's and b's alternate, the nth term is: $\frac{b+a}{2}+(-1)^n\frac{b-a}{2}$.

127. The hour and minute hands are 120° apart at 4 (or 8) o'clock, and then again between 5 and 6 (or 9 and 10), between 6 and 7 (or 10 and 11), etc., a total of 11 times with the minute hand before the hour hand, and 11 times in opposite order. Therefore, the positions of the minute hand that we seek are $\frac{60}{11}$ or $5\frac{5}{11}$ minutes apart. The hour hand can now be disregarded because the position of the second hand is determined by the position of the minute hand. Actually, only the $\frac{5}{11}$ is needed to determine that at $5\frac{5}{11}$ after the hour, the second hand is $\left(\frac{5}{11}\right)\cdot 60$ or $\frac{300}{11}$ after the hour. Therefore the positions, in minutes after the hour, of the minute and second hands that concern us are: $\left(\frac{60n}{11}, \frac{300n}{11}\right)$ for $n=0, 1, 2, \ldots 10$. Their difference, $\frac{240n}{11}$, can be a multiple of 20 (minutes) only for $n=0$, at which time the hands coincide. Thus it is impossible for the three hands to divide the face into three equal sectors.

Note:

This problem is referred to without solution by Dudeney. See [8], problem 63.

128. **First Solution**

If n is the number of wins of the last team, and d is the common difference of wins between successive teams, then since $18(1+2+\ldots+9)$, or 810 games are played,

$$n+(n+d)+(n+2d)+\ldots+(n+9d)=810.$$

This reduces to $2n+9d=162$. Now n will be a maximum when d is a minimum (but not zero, because there are no ties). The smallest integral value of d for which n is integral is 2; thus $n=72$.

Second Solution

There were 810 games won (see first solution). If there were an odd number of teams, and each team was the same number of games ahead of the following team, the middle team would have to play .500 ball. But with ten teams, the fifth and sixth teams must win 162 games between them, and play as close to .500 ball as possible. Therefore the fifth team must have won 82 and the sixth 80 games. With a common difference of two games, the lower teams must have won 78, 76, 74, 72 games.

129. **First Solution**

Of numbers of more than one digit, those made up entirely of 2's, 3's, 7's, or 8's cannot be squares because of their last digits.

Numbers made up entirely of 1's, 5's, or 9's cannot be squares because they are of the form $4n-1$. This can be shown by adding 1 in each case and observing that the last two digits form a multiple of 4.

Numbers made up entirely of 6's cannot be squares because they contain the factor 2 just once; whereas in a square a prime factor must be contained an even number of times. This leaves numbers made up entirely of 4's, which cannot be squares because $4 \cdot 111 \ldots$ is the product of a square by a non-square.

Note:

Every odd number is of the form $2a+1$. Its square is $4(a^2+a)+1$, which cannot be of the form $4n-1$.

Any number can be represented by $a+10b+100c+\ldots$, in which $100c+\ldots$ is at once seen to be divisible by 4. Hence if $a+10b$, that is, the number formed from the last two digits, is divisible by 4, the entire number is divisible by 4.

Second Solution

A table of squares shows that except for 00, 44 is the only ending of a square with two final digits that are alike, and that the numbers whose squares end in 44 must have final digits: 12, 38, 62, or 88. The square of a number ending in 12 is obtained thus:

$$\begin{array}{r}
\ldots x12 \\
\ldots x12 \\
\hline
\ldots (2x)\ 2\ 4 \\
\ldots\ \ 1\ \ 2\ \ \ \ \\
\ldots\ 2x\ \ \ \ \ \ \ \\
\ldots\ \ \ \ \ \ \ \ \ \ \ \ \\
\hline
(4x+1)\ 4\ 4
\end{array}$$

Since $4x+1$ cannot end in 4, the number $(\ldots 12)^2$ cannot end in all 4's.

The other cases are left to the reader.

130. The wording of the problem implies that we consider the surface to be fixed and the cylinder and ruler to be moving. But since we are considering only the relative positions of points on these three objects, we may suppose that the cylinder is fixed but able to rotate about O, and that the surface and ruler are moving. If the cylinder rotates and point A on the ruler moves x units relative to

O, then the surface must also move x units relative to O, but in the opposite direction. Thus, while the displacement between O and the surface is x units, that between A and the surface is $2x$ units, or twice as much.

Figure 79

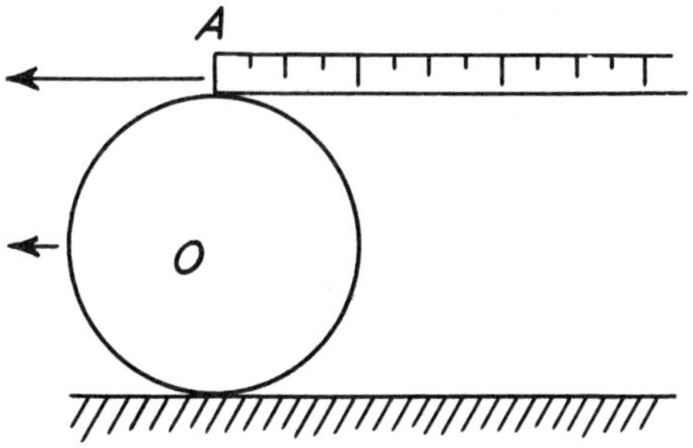

Note:

The next time you enter an elevator, note that one of its half doors slides twice as far as the other. This is accomplished by suspending one from points that correspond to O and the other from points that correspond to A on a gear mechanism.

131. The book is divided into three sections, each of which contains the same number of digits. The number of pages in the first section exceeds 9, and (if it does not also extend into the hundreds) they contain an odd number of digits. Thus, the second section is located partially in the tens and partially in the hundreds. Let x equal the number of pages in the first section beyond 9. Let y equal the number of pages in the second section beyond 99. Then $9+2x=2(90-x)+3y$; or $x=\frac{171+3y}{4}$. The least integral value of y which will make x integral is 3. Hence $(x, y)=(45, 3)$. Thus there are $9+45=54$ pages and $9+2(45)=99$ digits in the first section, and $(90-45)+3=48$ pages, and, as required, $2(45)+3(3)=99$ digits in the second section. We are now in the hundreds. The next 33 pages will contain 99 digits, as required. Thus the book contains a total of $54+48+33=135$ pages.

132. **(a)** n is even.

$y+z$ takes its smallest value when $y=z=1$, and $2x$ is then $n-2$. Thus we may let $x=1, 2, 3, \ldots, \frac{n-2}{2}$ We have then, $y+z=n-2, n-4, n-6, \ldots$ to which there are $n-3$,

$n-5$, $n-7$, . . . solutions in positive integers. Using the formula

$$S=\frac{N}{2}[2a+(N-1)d]$$

in which $N=\frac{n-2}{2}$ (the number of x's), $a=n-3$, $d=-2$, we obtain $\frac{n^2-4n+4}{4}$ solutions.

(b) n is odd.

Reasoning as in (a), $x=1, 2, 3, \ldots, \frac{n-3}{2}$; $y+z=n-2$, $n-4$, $n-6$, . . . to which there are $n-3$, $n-5$, $n-7$, . . . solutions. In the formula for S, we substitute $a=n-3$, $N=\frac{n-3}{2}$, $d=-2$, and obtain $\frac{n^2-4n+3}{4}$ solutions.

133. (a) If each column is numbered for one of the three coins as follows:

$$\begin{matrix} 1 & 2 & 3 \\ H & H & H \end{matrix}$$

we can indicate a solution to the problem by naming the column for each coin that is turned. One solution is: 3-2-3-1-2-3-2. These correspond to the changes: 1. *HHH;* 2. *HHT;* 3. *HTT;* 4. *HTH;* 5. *TTH;* 6. *THH;* 7. *THT;* 8. *TTT.*

If the numbers are permuted a new solution is obviously obtained. For example, 132 is a permutation of 123. Therefore interchanging 2 and 3 in the solution above results in the new solution: 2-3-2-1-3-2-3. There are thus 3!, or 6 solutions to this problem corresponding to the permutations: 123, 132, 213, 231, 312, 321.

(b) Let every arrangement of *H*'s and *T*'s be called even if it contains an even number of *T*'s, and odd otherwise. Step 1 (*HHHH*) is even, step 2 must be odd, step 3 even, etc. Step 16 should then be odd, since all even-numbered steps are odd. But step 16 has 4 *T*'s and is therefore even. Hence there is no possible solution for 4 coins, and in general for any even number of coins.

Note:

In general there will be $n!$ solutions for n coins if n is odd; no solutions if n is even.

134. Let x be the number of packets A, y the number of packets B, and z the number of packets C. There will then be $3x+2y+5z$ get-well cards, $x+3y+4z$ greeting cards, and $7x+y+2z$ anniversary cards. To form the new packets with one of each type of card, the number of each type of card must be the same. Let this number be represented by t. Then:

$$3x+2y+5z=t$$

$$x+3y+4z=t$$

$$7x+y+2z=t.$$

Solving for x, y, and z in terms of t we obtain:

$$x=\frac{2t}{21},\ y=\frac{5t}{21},\text{ and } z=\frac{t}{21}.$$

The smallest value of t that will make x, y, and z integral and positive is 21. Thus if we merge 2 packets of type A, 5 packets of type B, and 1 packet of type C, we will obtain:

$3\times 2+2\times 5+5\times 1=21$ get-well cards;

$1\times 2+3\times 5+4\times 1=21$ greeting cards; and

$7\times 2+1\times 5+2\times 1=21$ anniversary cards.

These can therefore be merged to form 21 new packets.

135.

Solution

Sin $12°$ sin $48°$ sin $54° = \frac{1}{8}$, if $(2 \sin 12° \cos 42°) \sin 54° = \frac{1}{4}$, if $(\sin 54° - \sin 30°) \sin 54° = \frac{1}{4}$, if $4 \sin^2 54° - 2 \sin 54° - 1 = 0$, if $\sin 54° = \frac{1+\sqrt{5}}{4}$. (Not $\frac{1-\sqrt{5}}{4}$, which is negative.) Now the latter can be verified in a number of ways. Then, since the steps are reversible, the given identity will have been proven.

First Method

Construct isosceles $\triangle ABC$ with $m\angle C=36°$, and $AC=CB=1$. Draw other lines as indicated in Fig. 80. Since $\triangle ABC \sim \triangle ADB$, $\frac{AB}{AC}=\frac{AD}{AB}$. Thus $\frac{x}{1}=\frac{1-x}{x}$, from which it follows that $x=\frac{-1+\sqrt{5}}{2}$.

Therefore from $\triangle CDE$, $\cos 36° = \sin 54° = \frac{1}{2x} = \frac{1+\sqrt{5}}{4}$.

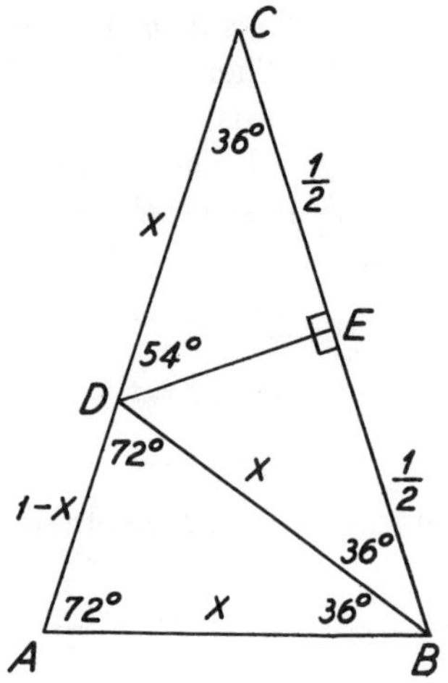

Figure 80

Since $\sin A \sin B = \frac{1}{2}[\cos (A-B) - \cos (A+B)]$, we obtain $\sin 12° \sin 48° \sin 54° = \frac{1}{2}\left[\frac{1+\sqrt{5}}{4} - \frac{1}{2}\right] \cdot \left[\frac{1+\sqrt{5}}{4}\right] = \frac{1}{8}$.

Second Method

Let $A = 18°$; then $\sin 2A = \cos 3A$, and $2 \sin A \cos A = 4 \cos^3 A - 3 \cos A$. Divide through by $\cos A$ (which does not equal zero), and substitute $1 - \sin^2 A$ for $\cos^2 A$. The resulting quadratic equation leads to $\sin 18° = \frac{-1+\sqrt{5}}{4}$. Substituting in $\sin 54° = \cos 36° = 1 - 2 \sin^2 18°$ yields $\sin 54° = \frac{1+\sqrt{5}}{4}$.

Third Method

Let one fifth root of 1 be $e = \cos 72° + i \sin 72°$, where $i = \sqrt{-1}$. Then $e^{-1} = \cos 72° - i \sin 72°$. Therefore $e + e^{-1} = 2 \cos 72° = f$, where f is chosen arbitrarily. Since e is a root of $x^5 - 1 = (x-1)(x^4 + x^3 + x^2 + x + 1) = 0$, and $e - 1$ does not equal zero, it follows that $e^4 + e^3 + e^2 + e + 1 = 0$. Divide by e^2: $e^2 + e + 1 + e^{-1} + e^{-2} = 0$. But $f^2 - 2 = e^2 + e^{-2}$. Therefore $f^2 + f - 1 = 0$ and $f = \frac{-1+\sqrt{5}}{2} = 2 \cos 72°$. Therefore $\sin 18° = \frac{-1+\sqrt{5}}{4}$. Now $\sin 54° = \sin 3(18°)$ leads to the same result as the other methods.

For other identities of this type see [17] and [29].

136. Since in $\triangle ABC$ $\cos A=\frac{b^2+c^2-a^2}{2bc}$ and $\sin A=\frac{a}{2R}$ (R=circumradius), then $\cot A=\frac{\cos A}{\sin A}=\frac{R(b^2+c^2-a^2)}{abc}$. Also $\cot B=\frac{R(c^2+a^2-b^2)}{abc}$ and $\cot C=\frac{R(a^2+b^2-c^2)}{abc}$. Therefore $\cot A+\cot B+\cot C=\frac{R(a^2+b^2+c^2)}{abc}$. But $\triangle=\frac{abc}{4R}$. Therefore the required result follows.

Let a circle be circumscribed about $\triangle ABC$, and let $\overline{CD}$ be an altitude and $\overline{CE}$ a circumdiameter. Draw $\overline{BE}$; then $\triangle CAD\sim\triangle CBE$, and $\frac{CA}{CE}=\frac{CD}{CB}$. Hence $ab=h_c\cdot 2R$ (h_c is the altitude to $\overline{CD}$). Therefore $\frac{ch_c}{2}=\triangle=\frac{abc}{4R}$.

We note also that $\frac{CB}{CE}=\sin E$ or, since $m\angle E=m\angle A$, $\sin A=\frac{a}{2R}$.

Figure 81

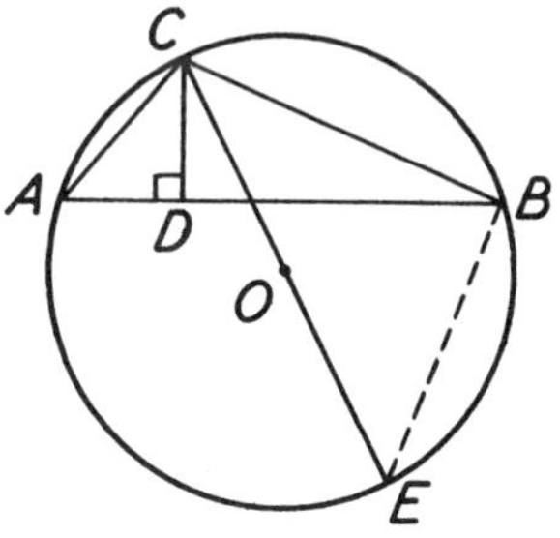

137. If n is the number of radios bought, then $\frac{432}{n}=a$, and $\frac{324}{n-18}=b$ are both integers. The factors of 324 are the terms of the product $(1+2+2^2)(1+3+3^2+3^3+3^4)$, viz. 1, 3, 9, 27, 81, 2, 6, 18, 54, 162, 4, 12, 36, 108, 324.

If $n-18$ is set equal to each of these in turn, we obtain as tentative values: $n=19$, 21, 27, 45, 99, 20, 24, 36, 72, 180, 22, 30, 54, 126, 324.

But n is a factor of 432. Therefore $n=27$, 24, 36, 72, 54. The values of (a, b) that correspond are, in order: (16, 36), (18, 54), (12, 18), (6, 6), (8, 9). In the case (6, 6), no profit is made. Hence the final answers are: $n=27$, 24, 36, 54.

138. If $N = p_1^{a_1} p_2^{a_2} \ldots p_r^{a_r}$, where $p_1, p_2, \ldots p_r$ are distinct primes, then the factors of N are the terms of the product $(1 + p_1 + p_1^2 + \ldots + p_1^{a_1})(1 + p_2 + p_2^2 + p_2^{a_2}) \ldots (1 + p_r + p_r^2 + \ldots + p_r^{a_r})$. There are $(a_1 + 1)(a_2 + 1) \ldots (a_r + 1)$ such factors.

Now let the required number $N = 2^{a_1} 3^{a_2} 5^{a_3} \ldots$ Then there are two possibilities:

(i) One $a_i = 5$ and all others $= 0$. For then $a_i + 1 = 6$, and all other factors $a_j + 1$ are 1's.

(ii) One $a_i = 1$ and another $a_j = 2$. For then, $(a_i + 1)(a_j + 1) = 2 \cdot 3 = 6$, and all other factors $a_j + 1$ are 1's.

In case (**i**) the only value of N between 200 and 300 is $3^5 = 243$. In case (**ii**) the only possible values of N are:

$2 \cdot 11^2 = 242$, $5 \cdot 7^2 = 245$, $11 \cdot 5^2 = 275$, $23 \cdot 3^2 = 207$,

$29 \cdot 3^2 = 261$, $31 \cdot 3^2 = 279$, $2^2 \cdot 53 = 212$, $2^2 \cdot 59 = 236$,

$2^2 \cdot 61 = 244$, $2^2 \cdot 67 = 268$, $2^2 \cdot 71 = 284$, $2^2 \cdot 73 = 292$.

139. For n squares on a side, the number of vertical walls is $(1 + 2 + 3 + \ldots + n) + n$. This is also the number of horizontal walls. Therefore the total number of walls, T, $= \dfrac{2n(n+1)}{2} + 2n = n^2 + 3n$. For $n = 16, 17$, $T = 304, 340$. Therefore the number of additional walls needed is $340 - 317$ or 23 walls.

140. Let x and y represent the 3-digit numbers HAT and BOX respectively. Then 9 $HATBOX = 4$ $BOXHAT$ becomes $9(1000\ x + y) = 4\ (1000\ y + x)$. Therefore $8996\ x = 3991\ y$, or dividing by 13, $692\ x = 307\ y$. Since 692 and 307 have no common factors, $x = 307$ and $y = 692$. Thus $HATBOX = 307{,}692$ and $BOXHAT = 692{,}307$.

APPENDICES

A. PROPERTIES OF PROPORTIONS

Theorem: *If* $\frac{a}{b}=\frac{c}{d}$, *then for any constants (except for division by zero),* $\frac{pa+qb}{ra+sb}=\frac{pc+qd}{rc+sd}$.

Proof

Let $\frac{a}{b}=\frac{c}{d}=t$; then $a=tb$ and $c=td$. Therefore $\frac{pa+qb}{ra+sb}=\frac{ptb+qb}{rtb+sb}=\frac{pt+q}{rt+s}$, and $\frac{pc+qd}{rc+sd}=\frac{ptd+qd}{rtd+sd}=\frac{pt+q}{rt+s}$. Hence, the theorem is true.

For $p=q=r=1$, and $s=-1$, the theorem becomes:

$$\text{If } \frac{a}{b}=\frac{c}{d}, \text{ then } \frac{a+b}{a-b}=\frac{c+d}{c-d};$$

in words: *numerator plus denominator over numerator minus denominator equals numerator plus denominator over numerator minus denominator.*

For $p=r=s=1$, and $q=0$, the theorem becomes:

$$\text{If } \frac{a}{b}=\frac{c}{d}, \text{ then } \frac{a}{a+b}=\frac{c}{c+d}; \text{ in words:}$$

numerator over numerator plus denominator equals numerator over numerator plus denominator. Reference: [18] Chapters 1 and 2.

B. CEVA'S THEOREM

If three concurrent lines $\overline{AG}$, $\overline{BG}$, $\overline{CG}$ *meet the opposite sides of* $\triangle ABC$ *at D, E, F respectively, then* $\frac{AF \cdot BD \cdot CE}{FB \cdot DC \cdot EA}=1$.

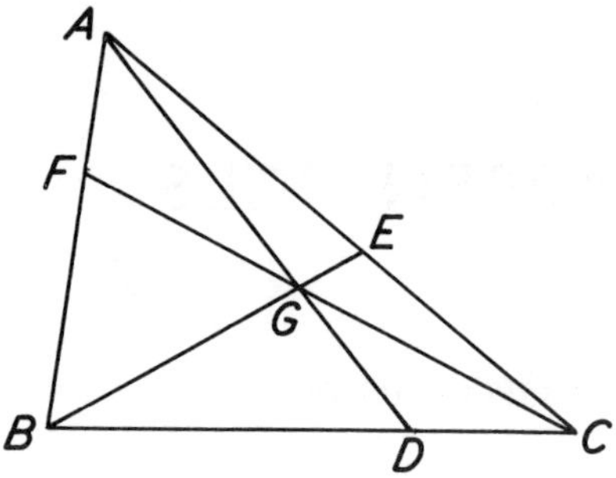

Figure 82

Proof

$$\frac{\text{Area } \triangle AGC}{\text{Area } \triangle BGC}=\frac{AF}{FB}$$

$$\frac{\text{Area } \triangle BGA}{\text{Area } \triangle AGC}=\frac{BD}{DC}$$

$$\frac{\text{Area } \triangle BGC}{\text{Area } \triangle BGA}=\frac{CE}{EA}$$

Multiplying left and right members, and canceling equal factors to to the left leads to the above result.
Reference: [5].

C. MENELAUS' THEOREM

If a line $\overleftrightarrow{FH}$ *intersects the sides* $\overline{AB}$, $\overline{BC}$, $\overline{CA}$ *of* $\triangle ABC$ *at F, H, E respectively, then* $\frac{AF \cdot BH \cdot CE}{FB \cdot HC \cdot EA}=1$. (This does not take account of directed line segments—a consideration found useful in advanced treatments of the subject.)

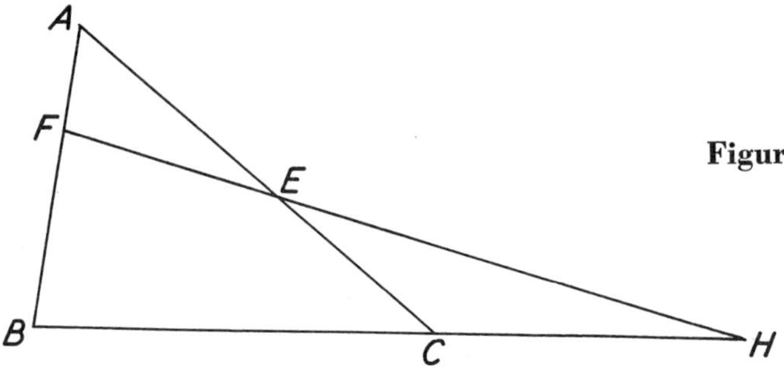

Figure 83

Proof

Draw perpendicular segments from B, A, and C to $\overleftrightarrow{HE}$, and represent their length by p, q, and r, respectively. Then $\frac{AF}{FB}=\frac{q}{p}$; $\frac{BH}{HC}=\frac{p}{r}$; $\frac{CE}{EA}=\frac{r}{q}$. Continue as in Ceva's Theorem.
Reference: [5].

D. HARMONIC SETS OF POINTS

If we combine the two figures above and apply the theorems in **B** and **C** we obtain:

$$\frac{AF\cdot BD\cdot CE}{FB\cdot DC\cdot EA}=\frac{AF\cdot BH\cdot CE}{FB\cdot HC\cdot EA}=1.\text{ Hence }\frac{BD}{DC}=\frac{BH}{HC}.$$

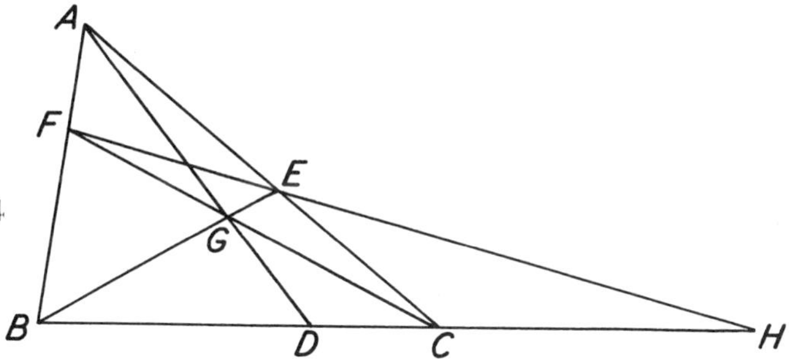

Figure 84

Thus D and H divide $\overline{BC}$ internally and externally in the same ratio. D and H are called Harmonic Conjugates with respect to B and C, and D, H, B, C are known as an harmonic set of points.

If in the last proportion, DC and HC are replaced by $BC-BD$ and $BH-BC$ respectively, it can be shown that

$$\frac{1}{BC}=\frac{1}{2}\left(\frac{1}{BH}+\frac{1}{BD}\right).$$

Compare this with the last paragraph in the solution to Problem 9 to see why the term "harmonic" is used.

Given three collinear points such as B, D, C, the harmonic conjugate H, of D, with respect to D can be readily constructed with straightedge alone by reconstructing Fig. 84. Thus, take any point A not on $\overline{BC}$ and draw $\overline{AB}$, $\overline{AD}$, $\overline{AC}$. Take any point G on $\overline{AD}$ and draw $\overrightarrow{BG}$ and $\overrightarrow{CG}$ intersecting $\overline{AC}$ and $\overline{AB}$ at E and F respectively.

$\overrightarrow{FE}$ will now intersect $\overrightarrow{BC}$ at H. The determination of the position of H is independent of the choices of points A and G.

If D is the midpoint of $\overline{BC}$, then its harmonic conjugate is the point at infinity, and $\overline{EF}$ will be parallel to $\overline{BC}$.

E. *CONCYCLIC POINTS*

Given four points of which no three are collinear, it is always possible to construct a circle which will pass through any three of them; but the circle may or may not pass through the fourth point. If the circle does pass through all four points they are referred to as Concyclic Points. The knowledge that four points are concyclic can be useful in proving that certain angles are right angles or are congruent or supplementary.

Here are two useful methods of proving that four points are concyclic:

(1) If $m\angle B = m\angle C$ or if $m\angle A = m\angle D$, then A, B, C, D are concyclic.

Proof

Draw a circle through A, B and C intersecting $\overline{BD}$ at D'. Then show that D' must coincide with D because $m\angle BD'C = m\angle BDC$. This can be done whether D' is inside or outside the circle.

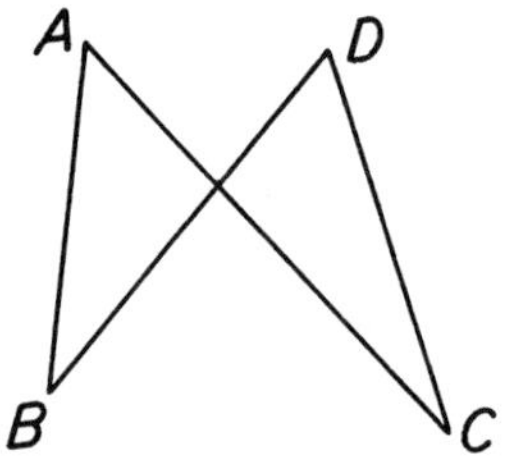

Figure 85

(2) If opposite angles of a quadrilateral are supplementary, the quadrilateral is concyclic.

For example if $\angle A$ is the supplement of $\angle C$, or if $\angle B$ is the supplement of $\angle D$, the quadrilateral is concyclic.

The proof is similar to that in (1).

Figure 86

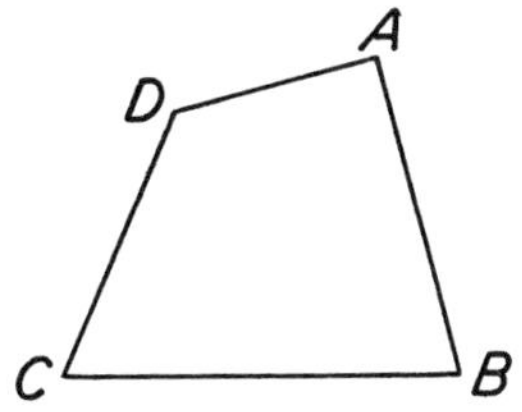

Note:

These are easy to remember because conversely, if A, B, C, D are concyclic, then in **(1)** $m\angle B = m\angle C$ and $m\angle A = m\angle D$; and in **(2)** $\angle A$ is the supplement of $\angle C$, and $\angle B$ is the supplement of $\angle D$.

As an application, consider $\triangle ABC$ and its altitudes $\overline{AD}$, $\overline{BE}$, and $\overline{CF}$. Prove that $m\angle 1 = m\angle 2$.

Figure 87

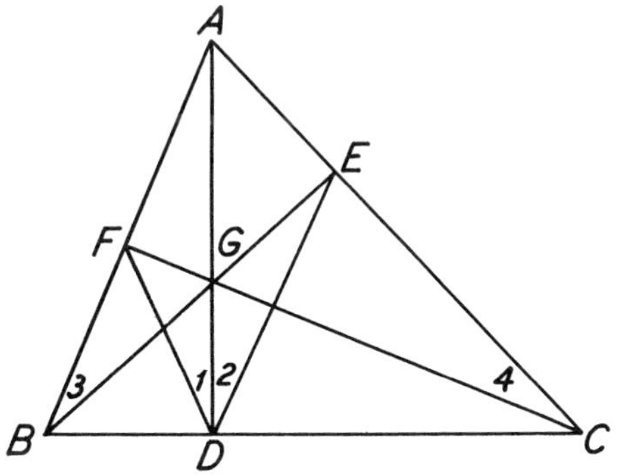

Proof: Since $\angle BFG$ is the supplement of $\angle BDG$, then B, F, G, D are concyclic. Hence $m\angle 1 = m\angle 3$; (both intercept the same arc). Since $m\angle CEG$ is the supplement of $\angle CDG$, C, E, F, D are concyclic. Hence $m\angle 2 = m\angle 4$. Now $m\angle BFC = m\angle BEC$, therefore B, F, E, C and concyclic. Hence $m\angle 3 = m\angle 4$. Therefore $m\angle 1 = m\angle 2$.

Another method is to compare the angles of $\triangle BFG$ and $\triangle GEC$.

Problem

Find as many sets of four concyclic points as you can in the above figure. Draw $\overline{FE}$ and find other pairs of equal angles by the above method.

F. PTOLEMY'S THEOREM

If a quadrilateral is inscribed in a circle, the product of the lengths of its diagonals is equal to the sum of the products of the lengths of pairs of its opposite sides.

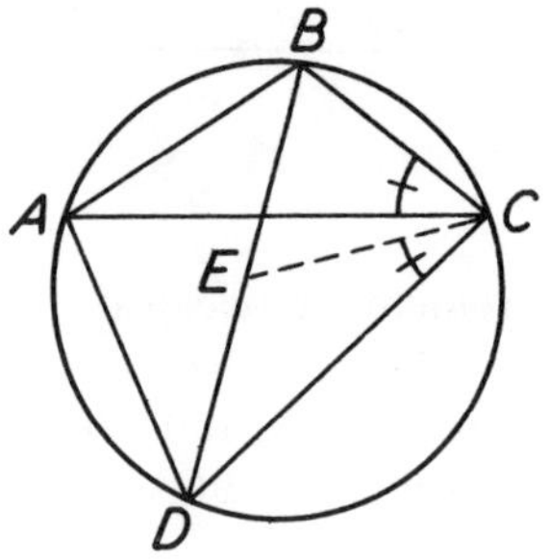

Figure 88

Proof: Draw $\overline{CE}$ so that $m\angle ECD = m\angle BCA$. Since $m\angle EDC = m\angle BAC$, $\triangle DEC \sim \triangle ABC$. Therefore $\frac{ED}{AB} = \frac{DC}{AC}$, and $AB \cdot DC = AC \cdot ED$ **(I)**. Also since $m\angle BCE = m\angle ACD$, and $m\angle CBD = m\angle CAD$, $\triangle BEC \sim \triangle ADC$.

Therefore $\frac{BE}{AD} = \frac{BC}{AC}$, and $AD \cdot BC = AC \cdot EB$ **(II)**. Add **I** and **II**: $AB \cdot DC + AD \cdot BC = AC(ED + EB) = AC \cdot BD$, which is the required result.

Applications

(1) If equilateral $\triangle ABC$ is inscribed in a circle, and D is any point on arc AC, then $DA + DC = DB$.

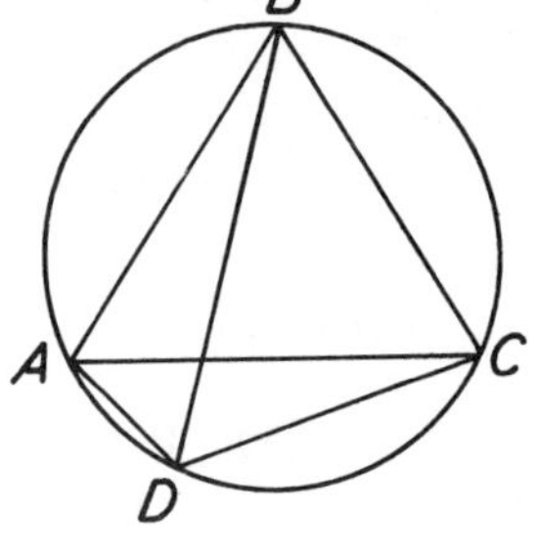

Figure 89

Proof: Let $AB=BC=CA=s$. By Ptolemy's Theorem: $AB \cdot DC + BC \cdot DA = AC \cdot DB$, or $s(DA)+s(DC)=s(DB)$, from which the required result follows.

(2) $ABED$ is a square on hypotenuse $\overline{AB}$ of right triangle ABC. If $AC=3$, $CB=4$, find CF, where F is the intersection of the diagonals $\overline{AE}$ and $\overline{BD}$.

Figure 90

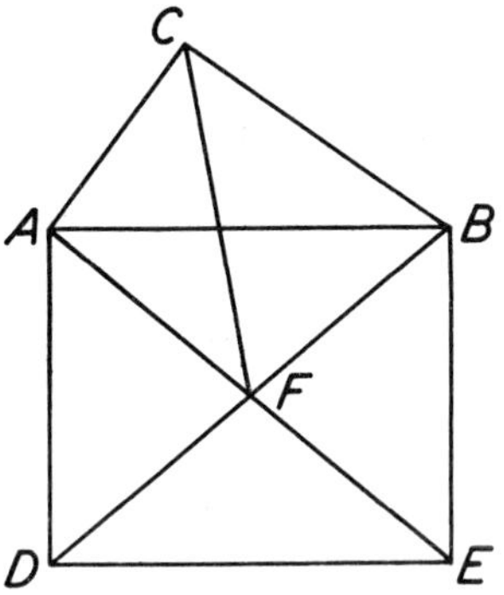

Solution: $m\angle ACB = m\angle AFB = 90°$. Therefore by Ptolemy's Theorem $CA \cdot FB + CB \cdot AF = AB \cdot CF$. Since $AB=5$, $AF=FB = \frac{5}{\sqrt{2}}$. Hence $3\left(\frac{5}{\sqrt{2}}\right)+4\left(\frac{5}{\sqrt{2}}\right)=5(CF)$. This gives the result: $CF=\frac{7}{\sqrt{2}}$.

BIBLIOGRAPHY

Problems of various levels of difficulty may be found in the following textbooks and collections of problems and puzzles. References to a number of them are made in the solutions to the problems.

1. BALL, WALTER W. R. *Mathematics Recreations and Essays.* (Revised by H. S. M. Coxeter.) New York: Macmillan, 1962.
2. BEILER, ALBERT H. *Recreations in the Theory of Numbers.* New York: Dover Publications, 1964.
3. CARROLL, LEWIS (C. L. Dodgson). *Mathematical Recreations of Lewis Carroll.* Part II: *Pillow Problems* and *A Tangled Tale.* New York: Dover Publications, 1959.
4. CHRYSTAL, GEORGE. *Textbook of Algebra,* Vols. 1 and 2. London: A. & C. Black, 1926. New York: Dover Publications, 1961.
5. COURT, NATHAN A. *College Geometry.* New York: Barnes & Noble, 1952.
6. COURT, NATHAN A. *Mathematics in Fun and in Earnest.* New York: New American Library of World Literature, 1961.
7. DEGRAZIA, JOSEPH. *Math is Fun.* New York: Emerson Books, 1961.
8. DUDENEY, HENRY E. *Amusements in Mathematics.* New York: Dover Publications, 1958.
9. DUDENEY, HENRY E. *The Canterbury Puzzles.* New York: Dover Publications, 1958.
10. FROHLICHSTEIN, JACK. *Mathematical Fun, Games, and Puzzles.* New York: Dover Publications, 1962.
11. GAMOW, GEORGE, and STERN, MARVIN. *Puzzle-Math.* New York: Viking Press, 1958.
12. GARDNER, MARTIN. *Mathematics, Magic and Mystery.* New York: Dover Publications, 1956.
13. GARDNER, MARTIN. *The Scientific American Book of Mathematical Puzzles and Diversions,* Vol. 1. New York: Simon and Schuster, 1959.

14. GARDNER, MARTIN. *The Scientific American Book of Mathematical Puzzles and Diversions,* Vol. 2. New York: Simon and Schuster, 1961.

15. GRAHAM, L. A. *Ingenious Mathematical Problems and Methods.* New York: Dover Publications, 1959.

16. GRIFFIN, HARRIET. *Elementary Theory of Numbers.* New York: McGraw-Hill Book Co., 1954.

17. HALL, HENRY S., and KNIGHT, SAMUEL R. *Elementary Trigonometry.* London: Macmillan, 1919.

18. HALL, HENRY S., and KNIGHT, SAMUEL R. *Higher Algebra.* London: Macmillan, 1924.

19. HALL, HENRY S., and KNIGHT, SAMUEL R. *Key to Higher Algebra.* London: Macmillan, 1932.

20. HEATH, ROYAL V. *Mathemagic: Magic, Puzzles, and Games with Numbers.* New York: Dover Publications, 1953.

21. HUNTER, JAMES A. H., and MADACHY, JOSEPH S. *Mathematical Diversions.* New York: D. Van Nostrand Co., 1964.

22. JACOBY, OSWALD, and BENSON, W. H. *Mathematics for Pleasure.* New York: McGraw-Hill Book Co., 1962.

23. JOHNSON, ROGER A. *Advanced Euclidean Geometry.* Formerly, *Modern Geometry.* (Edited by John W. Young.) New York: Dover Publications, 1960.

24. KEMENY, JOHN G.; SNELL, J. L.; and THOMPSON, G. L. *Introduction to Finite Mathematics.* Englewood Cliffs, N. J.: Prentice-Hall, 1957.

25. KENDALL, P. M. H., and THOMAS, G. M. *Mathematical Puzzles for the Connoisseur.* New York: Thomas Y. Crowell Co., 1962.

26. KLEIN, FELIX. *Famous Problems of Elementary Geometry.* (Translated by W. W. Beman and D. E. Smith.) New York: G. E. Stechert and Co., 1930. New York: Dover Publications, 1956.

27. KRAITCHIK, MAURICE. *Mathematical Recreations.* New York: Dover Publications, 1953.

28. LITTON INDUSTRIES, INC. *Problematic Recreations.* Vols. 1-5. Beverly Hills, Calif.: Litton Industries, n. d.

29. LONEY, SIDNEY L. *Plane Trigonometry.* London: Cambridge University Press, 1923.

30. Loyd, Sam. *Best Mathematical Puzzles of Sam Loyd*, Vols. 1 and 2. (Edited by M. Gardner.) New York: Dover Publications, 1959.

31. Lucy, R. M. *A Problem a Day.* Harmondsworth, Middlesex: Penguin Books, 1937.

32. Maxwell, Edwin A. *Fallacies in Mathematics.* London: Cambridge University Press, 1963.

33. Mott-Smith, Geoffrey. *Mathematical Puzzles for Beginners and Enthusiasts.* New York: Dover Publications, 1959.

34. Northrop, Eugene P. *Riddles in Mathematics: A Book of Paradoxes.* New York: D. Van Nostrand Co., 1944.

35. Petersen, J. *Methods and Theories for the Solution of Problems of Geometrical Constructions.* (Translated by S. Haagensen.) New York: G. E. Stechert and Co., 1923.

36. Phillips, Hubert ("Caliban"); Shovelton, S. T.; and Marshall, G. S. *Caliban's Problem Book.* London: T. De la Rue & Co., 1933.

37. Phillips, Hubert ("Caliban") *Hubert Phillips' Heptameron.* London: Eyre & Spottiswoode, 1954.

38. Phillips, Hubert ("Caliban"). *My Best Puzzles in Logic and Reasoning.* New York: Dover Publications, 1961.

39. Phillips, Hubert ("Caliban"). *My Best Puzzles in Mathematics.* New York: Dover Publications, 1961.

40. Pólya, György. *How to Solve It.* Princeton, N. J.: Princeton University Press, 1945. New York: Doubleday and Co., 1957.

41. Pólya, György. *Mathematical Discovery.* New York: John Wiley & Sons, 1962.

42. Pólya, György. *Mathematics and Plausible Reasoning*, Vols. 1 and 2. Princeton, N. J.: Princeton University Press, 1954.

43. Rapaport, Elvira, translator. *Hungarian Problem Books I* and *II.* Vols. 11 and 12, *New Mathematical Library.* New York: Random House, 1961.

44. Reinfeld, Don, and Rice, David. *101 Mathematical Puzzles and How to Solve Them.* New York: Sterling Publishing Co., 1960.

45. Salkind, Charles T., editor. *The Contest Problem Book.* Vol. 5, *New Mathematical Library.* New York: Random House, 1961.

46. Shklarsky, D. O.; Chentzov, N. N.; and Yaglom, I. M. *The USSR Olympiad Problem Book.* (Revised and edited by I. Sussman, translated by John Maykovich.) San Francisco: W. H. Freeman and Co., 1962.

47. Steinhaus, Hugo. *Mathematical Snapshots.* New York: Oxford University Press, 1960.

48. Steinhaus, Hugo. *One Hundred Problems in Elementary Mathematics.* Basic Books, 1964.

49. Williams, W. T., and Savage, G. H. *The Penguin Problems Book.* Harmondsworth, Middlesex: Penguin Books, 1940.

50. Wylie, Clarence R., Jr. *101 Puzzles in Thought and Logic.* Dover Publications, 1957.

51. Yaglom, I. M. *Geometric Transformations.* Vol. 8, *New Mathematical Library.* (Translated by Allen Shields.) New York: Random House, 1962.

52. Yates, Robert C. *The Trisection Problem.* Baton Rouge, La.: The Franklin Press, 1942.

53. Young, J. W. A., editor. *Monographs on Modern Mathematics.* London: Longmans Green and Co., 1924. New York: Dover Publications, 1955.